The Mysteries of
the Marco Polo Maps

The Mysteries of
the Marco Polo Maps

BENJAMIN B. OLSHIN

The University of Chicago Press   Chicago and London

BENJAMIN B. OLSHIN is associate professor of philosophy and the history and philosophy of science and technology at the University of the Arts, Philadelphia.

The University of Chicago Press, Chicago 60637
The University of Chicago Press, Ltd., London
© 2014 by The University of Chicago
All rights reserved. Published 2014.
Printed in the United States of America

23 22 21 20 19 18 17 16 15 14   1 2 3 4 5

ISBN-13: 978-0-226-14982-0 (cloth)
ISBN-13: 978-0-226-14996-7 (e-book)
DOI: 10.7208/chicago/9780226149967.001.0001

Library of Congress Cataloging-in-Publication Data

Olshin, Benjamin B., author.
  The mysteries of the Marco Polo maps / Benjamin B. Olshin.
    pages cm
  Includes bibliographical references and index.
    ISBN 978-0-226-14982-0 (cloth : alk. paper) —
ISBN 978-0-226-14996-7 (e-book)   1. Polo, Marco,
1254–1323?   2. Cartography—History.   I. Title.
  G370.P9O57 2014
    912.092—dc23
                                        2014003274

Contents

Illustrations

This story spans seven centuries and takes us from the farthest reaches of Asia to Italy and finally to the United States. But it begins modestly enough: in 1991, I came across a decades-old journal article by the late historian of cartography Leo Bagrow (1881–1957), an article that discussed an obscure and curious collection of old maps pertaining to the voyages of Marco Polo.[1] The article also spoke of a certain "Marcian F. Rossi," who had emigrated from Italy to America in the nineteenth century and possessed those maps. Rossi claimed that the maps and other documents had come down directly through his family line, and he traced a lineage that went all the way back to the thirteenth century.

A Complex History

Marcian Rossi's particular claims are found in a letter, dated 14 January 1948, which he had sent to Bagrow.[2] This was already some years after Rossi had begun, in 1933, to engage the Library of Congress, lending that institution some of the documents and sending photostat copies (an early form of photocopies) of others. Part of the letter is quoted by Bagrow in his article:

Marco Polo entrusted the maps to Admiral Rujerius Sanseverinus who had graduated the Nautical School at Amalfi. A number of centuries later his descendant Ruberth Sanseverinus married Elisabeth Feltro Della Rovere, Duchess of Urbino. In the year 1539 Julius Cesare de Rossi, Count of Bergeto, married Maddalena Feltro Della Rovere Sanseverinus to whom the Tenure of Cajiate was assigned; his grand-son Joseph de Rossi became Duke of Serre; this tenure was held till 1744, when it was transferred to the Duchy of Casale to [of?] Joseph de Rossi; his younger brother Antonio de Rossi was the father of Marciano de Rossi, my great grand-father.[3]

When I read this passage, the first mystery was this "Admiral Rujerius Sanseverinus": who was this admiral and what was his relationship to Marco Polo? Some investigations led to the conclusion that Marcian Rossi might have been referring to Ruggero Sanseverino, who was also known as "il Grande Ammiraglio" Ruggero (or "Ruggiero"), who died in 1305 and was also known as "Ruggero di Lauria," from the town of Lauria, where he was born. We also know that he was admiral of Sicily and Aragona, but did he in fact receive maps from Marco Polo?

There is no record of such a transaction, but the two men were contemporaries; Marco Polo returned from his trip to the east in 1295 and lived for many years after that. Several of the other claims that Marcian Rossi makes in his letter also seem to stand up to modern research. The "Ruberth Sanseverinus" he mentions seems to be Roberto Sanseverino (d. 1487/1488), and that figure indeed married Elisabetta (da) Montefeltro (d. Rome, 1503), who was the illegitimate daughter of the famous Federico, Duke of Urbino. The "Julius Cesare de Rossi" mentioned in the letter is probably Giulio Cesare de Rossi, who was assassinated in 1554. After Giulio Cesare de Rossi, the lineage becomes more difficult to trace. There was, as Marcian Rossi claims in his letter, a "Joseph de Rossi" (i.e., Giuseppe de Rossi), who indeed held the title of "Secondo duca di Casal di Principe." But the letter still led to a number of questions, and, of course, there were the maps and other documents themselves, which presented many puzzles.

I had read the article on what are now known as the "Marco Polo Maps" or the "Rossi Collection" when I was still a graduate student, as part of my general thesis research into the history of old maps and cartographic texts. It was only some eight years later, in the summer of 1999, that I looked at the article again, in the process of going through some old materials after moving back home from overseas. As I glanced through the article that second time, I began to wonder where these maps were and considered how difficult it might be to track them down after half a century. The article intrigued me because the connection between Marco Polo and maps has always been very uncertain, and here was a whole collection of them. I was also intrigued because no one seemed to have followed up on the case.

A number of telephone calls to libraries and municipal archives led to Marcian Rossi's son, Louis A. Rossi, and then to Louis's daughter Beverly Pendergraft (née Rossi), and finally to her son, Jeffrey R. Pendergraft. When I contacted him, he was rather surprised that someone had gone to the trouble to track him down but immediately engaged me in a discussion of the maps, asking what my interest was. It turned out that he actually had quite a number of materials in this Rossi Collection—there were maps and related texts in the collection, but also many pieces unrelated to Marco Polo, including old deeds, wills, and so forth, which also merit future study. I was surprised, indeed, that I was the first to have tackled this mystery since Bagrow; but at the same time I knew that at least among academics, many mysteries go unexplored because of a general aversion to risk among university researchers in the humanities.

Pendergraft noted that the maps had sat for some time in his home, and he was enthusiastic about the idea that these materials finally would be investigated. It also turned out that there were even more cartographic documents than the few that Bagrow had discussed in his 1948 article.

Pendergraft, a Houston energy executive, had inherited the documents through Marcian Rossi, his great-grandfather (see fig. 1). The family history concerning Marcian was intriguing, if incomplete. He had been born in a place called Baia e Latina, a town north of Naples, in 1869. He had emigrated from Italy as a teenager, heading to St. Louis and then San Francisco. Available records indicate that his full name was Filomeno Emanuele Marciano Rossi, but he variously went by Marciano, Marcian, and so on; he was also called "Uncle Phil" by some relatives in the United States. Marcian's father was named Luigi Rossi (1837–1898) and, according to another member of the family, may have served in the forces of Garibaldi. A photograph of him in uniform survives (see fig. 2).

Marcian seems to have led a relatively quiet life, working as a tailor, but he also pursued a range of interests, even writing a science fiction novel in English entitled *A Trip to Mars*, which was published in 1920. Family genealogy and history were a particular interest, and in addition to the study of the cartographic and other materials in his possession, Marcian also traded other old maps and documents—exchanges, unfortunately, of which there is no record. We know that he was a friend of Alberto Francisco Porta, an architect and professor at Santa Clara University, and they communicated at some length on their common interest in history and exploration. Marcian Rossi passed away in 1948

J. W. FISCHER,   9TH & FRANKLIN AVE.

1   Filomeno Emanuele Marciano Rossi. (Source: Richard and Victor Rossi.)

in San Jose, California, and, as noted, his collection of maps and other old documents came into the possession of the Pendergraft branch of the family.

Initial Studies

In 2001, I delivered a preliminary report to Pendergraft; this report comprised an analysis of the content of a number of the documents, along with some general notes on early cartography for reference.

Some time later, we together contacted the Library of Congress in Washington, DC. The Library of Congress was in possession of one of the maps, since Marcian Rossi had lent them to be studied back in the 1930s. It had been agreed at that time that the library could retain one of the maps, the so-called "Map with Ship." Extensive discussions began with Ronald Grim and John R. Hébert of the Geography and Map Division at the Library of Congress about what they knew of the history of the Rossi documents.[4] These discussions led to the discovery of

2   Luigi Rossi, at right, in military uniform. (Source: Richard and Victor Rossi.)

further background material concerning the maps and Marcian Rossi's communications with the Library of Congress.

It was during this period, then, that I carried out background research—looking at the claims of Marcian Rossi, and examining the correspondence between Marcian Rossi and the Library of Congress, correspondence that spanned several decades, from the 1930s through the 1950s. The file of these letters contained many interesting items: debates by various scholars about the maps, negotiations concerning the possible donation of the maps to the library, and even a letter from J. Edgar Hoover, former head of the FBI, who had been involved in some initial examination of the maps in the 1930s.

My research during this period also focused on other maps and geographic documents that might relate somehow to these Rossi documents, such as medieval and early Renaissance world maps, early discussions of explorations of the coastal areas of northeast Asia, and a Chinese legend about a journey in the ocean far beyond the Asian mainland. That research led to some of the initial clues discussed in this book, clues as to how the pieces of the story might fit together. The maps and documents demanded investigation into all kinds of areas, including genealogy, cartography, Italian history, and Chinese history, not to mention the Italian, Latin, Chinese, and Arabic languages. Among map aficionados, there was growing interest, and in 2006, I gave a well-received presentation entitled "From Northeast Asia to the Pacific Northwest: 'Marco Polo' Maps and Myths" at the 47th Annual Meeting of the Society for the History of Discoveries in Portland, Oregon. The following year, I published a scholarly article on my investigations, the first formal study of these Rossi documents since Bagrow's article from over a half century earlier.[5]

Despite these investigations, there were still many questions, and those questions came from the odd nature of the collection itself. When the famous Vinland Map first appeared in 1957, it too presented a curious historical puzzle to researchers—and is still the subject of debate as to its authenticity and meaning.[6] The map, later obtained by Yale University and still in its possession, shows what appear to be landmasses in North America discovered by Norse explorations prior to Columbus. If the map is genuine, it would have to be considered the earliest work showing part of the Americas. But controversy about the Vinland Map and its provenance continue with no resolution in sight.

But the Vinland Map was a single work; with the Rossi documents, there are over a dozen maps and related materials, with texts in Italian, Latin, Chinese, and Arabic; some of the documents are supposedly signed by Marco Polo's daughters, or other mysterious characters bearing the Polo name such as a mysterious "Lorenzo Polo." It was an odd combination of material, and it required a great deal of analysis.

The first question was where these maps came from. The immediate provenance was not in question: clearly, Marcian Rossi had possessed these maps for several decades, as evidenced by the correspondence to and from the Library of Congress. But from where had *he*, in turn, received the maps? According to his claims, the maps had come down to his family by way of Admiral Ruggero Sanseverino, a contemporary of Marco Polo, and then, several centuries later, from the Sanseverino family to the Rossi family. As noted earlier, in his letter to Bagrow, Marcian Rossi had claimed that "in the year 1539 Julius Cesare de Rossi, Count of Bergeto, married Maddalena Feltro Della Rovere Sanseverinus"; research proved that this claim was correct. It is not clear how Marcian Rossi had found out this information, but Italian genealogies are a strong possibility; it is also possible that he had a personal written family history in his possession.

The first step, then, was to put together a rough genealogy that would cover everyone from Admiral Ruggero Sanseverino to "Julius Cesare de Rossi" (Giulio Cesare de Rossi, 1519–1554). This led to all kinds of interesting connections, with various famous Italian families appearing in the eventual genealogical sketch: the Sanseverino clan, the Sforza family, and the Medicis. The pieces fit together, but the big questions remained at the top and the bottom of this lineage. At the top, there was the question of the relationship between Marco Polo and Admiral Ruggero Sanseverino; Marcian Rossi had claimed in his letter that "Marco Polo entrusted the maps to Admiral Rujerius Sanseverinus," but there has been no way to verify this statement—no historical account makes or confirms this claim.

At the bottom, there was the question of the lineage of Marcian himself; in his letter, he states that Giulio Cesare de Rossi had a grandson "Joseph de Rossi," who became Duke of Serre, and that "this tenure was held till 1744." There was a Giuseppe de Rossi ("Giuseppe" being the Italian equivalent of "Joseph") who was a direct descendent of Giulio Cesare de Rossi, but he came some five generations after Giulio. This Giuseppe de Rossi was the Second Duke of Casal di Principe and died in 1779. His father, Gerardo de Rossi, had been the Fifth Duke of Serre, "Signore di Persano," and First Duke of Casal di Principe. Sources tell us that Giuseppe ceded both Serre and Persano, in exchange for Casal di Principe.

So the description in Marcian Rossi's letter differs in some respects, perhaps, from other accounts of the Rossi family. But the critical element was Marcian's statement that Giuseppe de Rossi had a younger brother, Antonio de Rossi, who "was the father of Marciano de Rossi, my great grand-father." These later generations were very difficult to verify, since there was the lack of clarity concerning the identity of Giuseppe de Rossi, and a lack of evidence concerning both Antonio de Rossi and Marciano de Rossi. Tracing the Rossi clan was particularly difficult due to the fact that the name is such a common one. But Marcian Rossi's letter was intriguing—in that it connected the distant past with much more recent family history, and that it was also correct in much of what it outlined. A break also came when I found two living members of the Rossi family, brothers in the United States, who supplied critically important information about Marcian Rossi and his relatives that at least made the family picture more complete.[7]

While the genealogical pieces were being fitted together, it was time to examine the maps and documents themselves. As far as we know, Marco Polo left no maps related to his travels. Indeed, some have expressed doubt as to whether Marco Polo ever traveled at all, since the narrative about his trip through Asia is secondhand, apparently penned by his cellmate, Rustichello da Pisa, to whom Polo dictated his oral account. Moreover, the narrative seems to leave out key elements of Chinese culture, such as the drinking of tea, and makes no mention of the Great Wall.[8] But the narrative itself, often called *Il Milione* (and known by the English title *The Travels of Marco Polo*), persists, and we have the fact that this text played a key role for several centuries in both map-making and exploration.[9] Toponyms from Marco Polo appear on maps even as late as the sixteenth century; Columbus himself owned and annotated a copy of the narrative and was influenced by the work, especially Polo's description of Japan.[10] Yet Marco Polo himself traveled in a period that had yet to see the development of sophisticated empirical cartography, and most of the maps that survive from this time—the late thirteenth century—are medieval world maps. There is, of course, also a tradition of early sea charts, but the Rossi documents appear to be unrelated to those types of works.

Perspectives and Questions

It was only with the publication of the article by Bagrow in 1948 that there appeared the possibility of finding a very close connection among the travels of Marco Polo, the narrative of *Il Milione*, and cartography from the actual period of his travels, that is, the late thirteenth and early fourteenth century. In his article, Bagrow reproduced a number of the maps—maps that Rossi claimed descended directly from the family of Marco Polo.

As noted earlier, Rossi, in fact, had submitted photostat copies of some of the maps to the Library of Congress some time before the Bagrow article, in 1933. The Library of Congress ended up receiving one of these maps as a gift from Rossi. That piece is the "Map with Ship," so called because of a small illustration of a ship in the document. The map is still in possession of the Library of Congress. In 1933, the library issued the first of a series of very brief notices on these maps, but no complete studies or publications were carried out.[11] In late 1933 and early 1934, there appeared short articles on one of the "Marco Polo Maps" in the *New York Times*, and in newspapers in Chicago and Minneapolis. A short mention of the maps is also found in 1937.[12]

Between 1933 and 1949, there was an exchange of letters between Rossi, Colonel Lawrence Martin, a few scholars, and others concerning these materials.[13] Most of the correspondence deals with the loan of the materials to the Geography and Map Division of the Library of Congress, as well as various conjectures concerning their interpretation. However, the letters also reveal that the division attempted to obtain some more in-depth input from scholars as to the content and authenticity of some of the maps.

There were many questions and doubts. A letter dated 17 May 1937 from David Magie (1877–1960), a classicist at Princeton, provides an attempted translation from one of the maps (the "Pantect Map"), and notes the following: "As to the genuineness of the document, I can form no opinion. I have shown it to the paleographer Lowe,[14] who says that he never saw such a hand-writing, and to Chalfont Robinson, who is the Curator of Medieval Manuscripts in the University. He looked through reproductions of about twenty 13th century manuscripts and found no writing similar to it, and also expressed great skepticism about the map . . . being too good for 1297." Two days later, on 19 May, Colonel Martin sent a reply to Professor Magie, thanking him but including the interesting following line: "Surely no modern forger would produce anything so illogical and mixed up if he were of a mind to deceive us." I will return to the question of authenticity in the final chapter of this book; in the meantime, Martin's suggestion is a sober one—why indeed would a forger produce something so *unconnected* to other maps of the period? In the same letter, Martin added: "I am not inclined to be disturbed because your colleague considers the map to be too good for the year 1297. No map . . . it seems to me, is impossible because we happen not to have yet seen one similar to it." Not an odd idea—a number of well-authenticated early maps display unique and unusual representations of the lands and seas: the early fifteenth-century Albertin de Virga map, for example; the Portolano Laurenziano-Gaddiano; and an anonymous manuscript world map of about 1530 in the Vatican. The first work is peculiar in that it includes a very large landmass extending northwast from Scandinavia, while the second is noted for its delineation of the coastline of Africa in a rather modern form, long before the Portuguese had rounded the Cape of Good Hope. The anonymous map in the Vatican includes a strange depiction of a large, ringed antipodal continent.[15]

This set of letters reveals another scholar's opinion, as well—that of Dana B. Durand, the well-known historian of science.[16] In a five-page memo from 16 May 1933 to the Library of Congress, Durand commented on one of the maps that he had seen from the collection.[17] In this memo, he states: "As limiting dates for the execution of the maps one would be safe in fixing 1400 and 1700." For Durand, important factors in looking at the maps were their depictions of Asia—which are indeed unusual, as we shall see—and the paleography. As to the latter, though, he states, "I am unable to affirm with complete finality, either from my own experience or the opinion of colleagues, but I feel, nevertheless that the general style of the script is as late as the 17th century."

Durand did not necessarily feel, though, that the maps were therefore simply fabrications from this period—rather, he implies that they were copies of earlier material, with additions and edits. For example, in a note in this same memo on one of the maps, he says, "The string of islands . . . on the *recto* looks as though it might have been sketched in later."

Durand seems to have remained uncertain on the matter, but had doubts that the maps could be from the period of Marco Polo per se. In a subsequent letter just a short time later (22 May 1933), he suggests the following: "The original parchment may have contained on it simply the outline of Europe and Asia on the *recto*. Coming into the hands of some late Venetian, possibly *ca.* 1650 to 1750, it may have roused some antiquarian interest. Adding to it the sketchy outlines to the east of Asia . . . he then proceeded to correct or improve it." Durand concludes that "the ultimate origin of the work" might be "a late hoax inspired by the discovery of an earlier sketch," adding that nonetheless "as a critical problem . . . [this] is of considerable interest." In 1936, a newspaper article summarized the situation as follows:

Prof. Dana Durand of Harvard University states he believes the documents shown him to be authentic. "The possibility of a forgery seems very doubtful." As has already been stated, Colonel Martin also considers them genuine, and so do a number of his colleagues.

There are however, world authorities on the other side. A cablegram from Europe reads, "British Museum authorities declare Polo maps spurious."[18]

One might be tempted to be sympathetic to this latter opinion. But this same newspaper article goes on to quote the British Museum's "keeper of the maps" as uttering the following rather absurd rationale: "The language contains bad Latin which would not have been employed by so good a classical scholar as Marco Polo."

Several decades later, there is a letter penned by Christopher de Hamel, who was at the time an assistant director at Southeby's and a specialist in medieval illuminated manuscripts.[19] In this letter, from 26 October 1979, he comments that the documents "present numerous problems," and expresses some of the same sentiments as Durand, noting, for example, that these materials may be "eighteenth-century copies of original items which have now been lost and survive only in the present copies," or that "parts (or all) of the documents are original, but that they have been re-inked and annotated in the eighteenth century."

He goes on to say that the idea that the documents are "eighteenth-century copies of original items" indeed "merits careful investigation." On a further positive note, he says that as to their being "outright forgeries" of some kind, "I do not think this is what they are." De Hamel committed himself further, in fact, stating that the maps could be described as "an exciting and enigmatic archive" and that Southeby's "could sell them well—even if we admit that they are probably eighteenth-century." He adds, "Assuming that we can argue that there is at least a probability that the maps are based on earlier lost exemplars, the collection is likely to be worth in the thousands of pounds."

The maps, though, apparently never appeared at auction. This leaves us—fortunately, we might say—with the same collection that puzzled these scholars. At the same time, as this book will demonstrate, there are both more questions than those faced by these correspondents *and* more evidence to work from. Here, for example, I have deciphered some of the Latin that perplexed Durand and found more detailed connections concerning some of the Chinese content.

Doubts and Possibilities

A careful—although incomplete—survey of the "Marco Polo" materials was carried out by William J. Wilson of the Library of Congress, around 1950.[20] Wilson prepared another report, of greater length, in 1953; this was entitled *The Rossi Collection of Manuscript Maps and Documents* and gives a good overview of the materials.[21] However, the report was intended only for use by the Library of Congress staff and was never published.

In his report, Wilson highlighted especially the question of authenticity. He recounted the dismissal by the British Museum of the maps as "spurious" and recounted how Italian experts agreed. In particular, the problem with these materials was that the "writing among them was [not] as early, by some centuries, as it professed to be."[22] In the first part of the report, Wilson went on to conclude that elements of the handwriting, capitalization, and punctuation all indicated that these materials were neither medieval nor Renaissance.[23] He stated: "These manuscripts are far later than they profess to be."[24] Yet at the same time, Wilson said: "And yet the case, though strong, is not logically complete. May not these manuscripts be late copies of authentic originals? . . . Could, indeed, the paleographical evidence be so interpreted that we might regard them, not as 19th or 20th century copies, but as 18th or even 17th century copies of 13th or 14th century originals? This is a tempting theory that has been plausibly presented by a succession of eminent scholars from 1933 down to 1948. No fair-minded investigators can possibly ignore it."[25] The complexity of the authenticity question is captured in a quote Wilson gives from a letter from Durand to Martin dated 18 October 1933. Durand is writing about what is labeled here the "Pantect Map": "The bulk of the outline on the *recto* may have been drawn as far back as the 14th century. If so, probably the Transerica region was added later. The Marco Polo allusion may be authentic, but if so is probably copied from some source other than the map itself. It would be doubtful then just how to classify the map, not exactly a forgery, nor yet exactly what it claims to be."[26] Wilson goes on to quote the article by Bagrow, where that writer states that the maps are not fakes but "more or less modernized copies."[27] Throughout the report, Wilson's tone is skeptical, and at the end of his review of the opinions of others such as Durand, he says that this collection "puts Marco Polo in possession of special information, not then generally known and not recorded in his book, about northeast Asia and the Alaskan region." At the conclusion of the report as a whole, he goes so far as to say that, compared to the other speculations on offer, a theory of forgery "would have more to commend it."[28]

As every investigator knows, extraordinary claims require extraordinary evidence, and the cartographic depictions here indeed make some extraordinary claims, as Wilson recognized. But while Wilson's study was intensive, it was not complete. This book examines many other aspects of the materials not investigated by Wilson, aspects that may lend further clues to understanding the materials. The examination includes full translations and analysis of the peculiar passages in Latin, a study of the Ptolemaic depictions in some of the maps, a look at the strange mention of Antilla, and further insights into a "Chinese connection." Wilson was right to be ambivalent—but also because there is much more to explore here.

Briefly looking back to the history of these documents, it is interesting that even the FBI became involved in the "Marco Polo Maps" mystery. We know that the "F.B.I took photographs using ultraviolet and special filter techniques to enhance the faded ink."[29] There is actually a letter in a file in the Geography and Map Division of the Library of Congress from J. Edgar Hoover. He delivered a brief report on one of the maps to the Library of Congress in 1944. However, his report provides no details as to the map's possible interpretation, and no conclusions as to the question of authenticity.

The "Marco Polo" documents received a brief notice in 1934 from Colonel Martin.[30] Bagrow's article was the first reasonably comprehensive scholarly analysis of the maps, but his work has some inaccuracies and omissions. For example, the Chinese text on the maps is not really examined in depth, nor is much attention given to the toponym Antilla. Much later, long after the Wilson report, the Library of Congress put together a short summary of the collection and related materials, but this was never widely circulated.[31] In terms of reviewing the actual documents, that summary examines the "Map with Ship," but then only giving a brief description of its contents. An interesting but also incomplete study of a number of the Rossi maps appeared in a rather obscure collection of Asian studies in 1965.[32]

This book is designed to introduce the reader to these maps, and to provide a detailed look at these fascinating materials: their depictions of lands and seas, how they relate to existing cartographic works and historical narratives, and the questions they raise. As we explore these items, we will find ourselves encountering many areas of knowledge, from early European exploration of the oceans to ancient Chinese legends. Much remains to be done with the Rossi documents, but this book lays a foundation for a deeper understanding of the material, and at the same time engages the reader in a genuine historical mystery.

ONE

The "Marco Polo Maps" and the Polo Family

The pieces that make up the "Marco Polo" collection of maps include a variety of intriguing images and texts. Not only do they include representations of lands and seas— along with what appear to be historical narratives—but they are also connected to one another through various cross-references and persons mentioned in more than one document. Before examining the materials in detail, we should take a quick look at exactly what the collection includes.

A Look at the Maps

There are fourteen documents in the Rossi Collection that concern Marco Polo and his travels. There are a number of connections between the various maps and texts.[1] A brief inventory is given below; the documents are categorized in terms of what names (e.g., "Bellela Polo") appear on the documents, and what cartographic images or geographic regions they treat.[2]

Document 1 ("Sirdomap Map"; see pl. 1)
A map of northeastern Asia with toponyms.

Document 2 ("Sirdomap Text"; see pl. 2)
A short text in "Arabic" lettering, followed by a brief Italian text,
    with the year "1267."

Document 3 ("Bellela Polo Chronicle"; see pl. 3)
A text in Italian concerning Marco Polo and Sirdomap.

Document 4 ("Map with Ship"; see pl. 4)
A map of eastern Asia, along with a picture of a sailing vessel.

Document 5 ("Pantect Map"; see pl. 5)
A map of eastern Asia, with an attached accompanying text.

Document 6 ("Fantina Polo Map 1"; see pl. 6)
A map covering Europe, North Africa, and Asia, with a "longitude-latitude" grid and a series of place-names.

Document 7 ("Fantina Polo Map 2"; see pl. 7)
A map depicting East Asia, a strait, and a peninsula with a chain of islands.

Document 8 ("Moreta Polo Map 1"; see pl. 8)
A map covering Europe, North Africa, and Asia, with a "longitude-latitude" grid.

Document 9 ("Lorenzo Polo Chronicle"; see pl. 9)
A text concerning the Polo family and concerning manuscripts left by "Rugerio Sanseverino."

Document 10 ("Map of the New World"; see pls. 11a and 11b)
On the recto, a map of Europe, North Africa, and North and South America; on the verso, there is a text mentioning Antilla and the explorer Hernando Cortez.

Document 11 ("Columbus Map"; see pl. 12)
A map of the New World, with a brief text.

Document 12 ("Spinola Chronicle"; see pl. 10)
A text in the form of a letter to "Elisabetta Feltro della Rovere Sanseverino" and signed "Guido Spinola."

Document 13 ("Keynote to Pantect Map"; see fig. 3)—*missing*
A text describing a voyage by Marco Polo to a chain of islands and a large peninsula in the Far East; this document is reproduced and discussed by Bagrow in his article of 1948, but is now missing from the collection.[3]

Document 14 ("Moreta Polo Map 2"; see figs. 4a and 4b)—*missing*
On the recto: a map of Asia, with an oval cartouche containing an inscription in

3 Document 13 ("Keynote to Pantect Map"), recto and verso. (Source: Leo Bagrow, "The Maps from the Home Archives of the Descendants of a Friend of Marco Polo," *Imago Mundi* 5 [1948]: 3–13.)

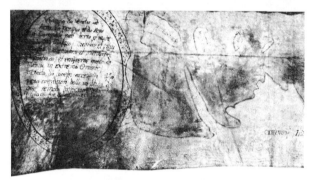

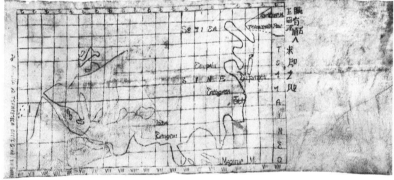

4a   Document 14 ("Moreta Polo Map 2"), recto (full and close-up views). (Source: Leo Bagrow, "The Maps from the Home Archives of the Descendants of a Friend of Marco Polo," *Imago Mundi* 5 [1948]: 3–13.)

4b   Document 14 ("Moreta Polo Map 2"), verso. (Source: Leo Bagrow, "The Maps from the Home Archives of the Descendants of a Friend of Marco Polo," *Imago Mundi* 5 [1948]: 3–13.)

Italian; on the verso: a map covering Europe, North Africa, and Asia, with a "longitude-latitude" grid. This document is discussed by Bagrow in his article but is now missing from the collection.[4]

The documents are all on parchment, with the maps and text done in ink. Most of the texts are in Italian; there are also shorter passages in Latin, Chinese, and Arabic. Most of the writing is very clear. However, there appears to have been erasing of text in some cases, with some underwriting slightly visible—this is primarily so with the "Map with Ship" (pl. 4) and the "Columbus Map" (pl. 12). In at least one case, there has been slight erasing and modification of text, as we will see in the discussion of the "Fantina Polo Map 1" (pl. 6), below. As noted in this list, two of the documents are missing, and the current owner, Jeffrey R. Pendergraft, believes that there may have been others in the collection at one time that are also now missing.

Initial Details and Clues

Looking at our list, we see immediately that there are three peculiar documents referring to a Syrian mariner named "Biaxio Sirdomap." The first is a map of northeastern Asia with toponyms referred to by the Roman numerals I through V, and with the toponyms written in some kind of Arabic text (see pl. 1). The second is a short text in this same script, followed by an Italian text on the same page (pl. 2). The third document is a text, apparently written by Marco Polo's daughter, Bellela Polo, about her father and his encounter with Sirdomap in the Far East (pl. 3). This document is known as the "Bellela Polo Chronicle" and is one of several pieces in the collection directly citing Marco Polo's daughters.

The "Map with Ship," currently in possession of the Library of Congress, has a rendering of the eastern regions of Asia, along with a picture of a sailing vessel (see pl. 4). The document appears as if it has been written on at different times, and there are bits of writing that have been scratched away. A complete discussion of this map is found in chapter 3 here.

The "Pantect Map" derives its name from a note in its right-hand margin that reads "Pantect De Praefectoria Potestate Interpret / in lat / Domus de Sanseverinus Urvinum." It, too, is a map of Asia, with the place-names taken from the ancient geography of Ptolemy. An inscription in Italian alongside the map claims that an accompanying Latin text above it is a translation from the original Tartar of the famed "Golden Tablet" given to Marco Polo by the great Kublai Khan (1215–1294). The map was folded and stored in a parchment envelope.

In 1934, Marcian Rossi sent a document from his collection to the Library of Congress. Now missing, this item survives only in the reproduction by Bagrow in his 1948 article. He calls it the "Keynote to the Pantect Map," although it does not seem to have any strong connection to that map. It is a text, in two parts, that describes a voyage by Marco Polo to a chain of islands and a large peninsula.

History tells us very little about Marco Polo's daughters, so it is rather suprising that a whole series of documents in the collection apparently contain their comments about their father's travels. The names of all Polo's daughters—Fantina, Bellela, and Moretta—appear on these maps and in these texts, which relate and refer to one another in various ways. The first document—the "Fantina Polo Map 1" (see pl. 6)—is a map covering Europe, northern Africa, and Asia, set in a kind of "longitude-latitude" grid, with place-names referred to by a series of Roman numerals. The map is signed "Fantina Polo" with the year "1329." Very similar gridded configurations are found in two other maps in this series. Another map, the "Fantina Polo Map 2" (see pl. 7), is also signed by Fantina Polo. It bears the same date and takes the form of an oval cartouche. The text below the map is a variant of that found in the first Fantina Polo map, just described.

The name of Moreta Polo—Marco Polo's youngest daughter—appears on several maps in this series. The first map, "Moreta Polo Map 1" (see pl. 8), displays the same kind of gridded configuration found in the "Fantina Polo Map 1." Another piece, the "Moreta Polo Map 2" (see figs. 4a and 4b), actually has maps on two sides. On one side, we find a map covering parts of South and East Asia and an oval cartouche, and on the reverse we see a map covering Europe, northern Africa, and Asia, again with a "longitude-latitude" grid. Both of these maps contain another surprise: Chinese characters written in the margins.

The "Lorenzo Polo Chronicle" (see pl. 9) presents us with more family names. This document is a long text discussing the Polo family and manuscripts left by a certain "Rugerio Sanseverino." The names of both Fantina and Moreta again appear, as does a certain "Lorenzo Polo," whose identity is uncertain. The relationship among all the documents in the Rossi Collection can be quite complex, and this "Lorenzo Polo Chronicle" adds to the complexity, presenting connections to a number of the other maps and texts.

The next document, the "Map of the New World" (see pls. 11a and

11b), in a manner similar to some of the other pieces of the collection, gives us both the familiar and peculiar. On the recto side of this document, there is a map of Europe, parts of northern Africa, and North and South America; the peculiarity is that the Americas are labeled "Columbia Septentrionalis" and "Columbia Meriodionalis," terms that seem not to be found in any other historical sources. Below the map, there is an accompanying text in the form of a letter that is addressed to "Elisabetta Feltro della Rovere Sanseverino," and that is signed "Guido Spinola" and labeled "Cagliari, 20 October 1524." On the verso is a text mentioning the mysterious island of Antilla, and a reference to the famed conquistador Hernando Cortez (1485–1547).

The final map under study here is the "Columbus Map" (see pl. 12); this is a map of the New World, with a brief text that is signed with the surname Sanseverino, which appears several times in these documents. The text mentions voyages by Columbus "and others" and notes that the map includes "islands and terra firma explored up to the year 1535." The collection also includes a text, known as the "Spinola Chronicle" (see pl. 10), which is in the form of a letter to "Elisabetta Feltro della Rovere Sanseverino" and signed by again by the mysterious Guido Spinola.

It is intriguing to think about the various people who have handled these documents over the centuries. Who examined them, read them, pondered their contents, and made their own mark? We are given a few clues by the documents themselves. The "Fantina Polo Map 2," a map of the northeastern reaches of Asia, set in oval outline and signed "Fantina Polo," has a small, oval-shaped tab attached to it at the bottom. On the tab is written: "diana bonacolsin da Verona." The last name was probably Bonacolsino or Bonacolsini, with the final vowel dropped. Furthermore, the -ini ending is a diminutive, so we can view the name as a variant of Bonacolsi—this was a famous Italian family that controlled several northern cities, including Mantua and Modena in the early thirteenth century.

Another Rossi document, the "Bellela Polo Chronicle" (see pl. 3), also has one of these attached, oval-shaped tabs. On this tab, we find a different name: Marta Veniero da Padova. The name Veniero is from the Venetian form of Venerio. Here, we might note a Venetian connection; indeed, Venerio was the surname of one of Venice's noble families.

Again, the daughters of Marco Polo play a key role in the mystery of these maps and texts, since their names—Bellela Polo, Moreta Polo, and Fantina Polo—are found on a number of the documents. What do we know about these women, and what do historical sources tell us about any connection between them and the voyages of their father? A reading of the maps seems to reveal that that the daughters were recording or preserving information about Asia that had been brought back by their father. But there are many questions brought up by such a scenario.

Nothing in the surviving manuscripts of the Polo narrative indicates that Marco Polo transmitted information about his journey to his daughters or left any behind for them. The Venetian historian, diplomat, geographer, and writer Giovanni Battista Ramusio (1485–1557), who put together an edition of the Marco Polo narrative and wrote a somewhat confusing summary of the Polo family, also does not mention any such transaction.[5] In one of the Rossi documents here, however, Bellela Polo clearly states: "So that the noble readers may find more delight concerning the Kingdom of Women in China and the Far East, my father Marco Polo wishes me, Bellela, to show this world map which he obtained from the pilot Biaxio Sirdomap." In another document, we read: "[The] dear princesses, duchesses, marchionesses, countesses, and all the ladies whom it would please to hear about the Realm of Women in China and the Far East, will be able to read that which my father recounted, and I, Bellela, have written from that account." Neither of the stories that Bellela then goes on to recount appears in the Polo narrative in the forms in which we know them today.

We find the name of Moreta Polo in two maps in this collection. In one document, the "Moreta Polo Map 1" (see pl. 8), we have a rendering of Europe, North Africa, and Asia, along with an inscription. It is signed "Moreta Polo," with the year "1338." We do not know when Moreta Polo was born, but we know that she died in 1375. Her father died in 1324, so this map must have been created after his death. Alternatively, Moreta simply could have added the inscription to the map—a map that would have been rendered at some earlier time by her father, or rendered by her based on notes from her father. The so-called "Moreta Polo Map 2" (see figs. 4a and 4b) has, on the recto side, a map of Asia, with an oval cartouche containing an inscription in Italian. On the verso, we see a map covering Europe, the northern part of Africa, and Asia. The map on the recto is signed as follows: "Morett[a] Polo. Venetia." There is an indication, as we found in the case of Bellela Polo, that this information was given to Moreta Polo by her father. The full inscription on the recto, in fact, ends with the phrase "tracto (d)a le lettere di mio p(ad)re," that is, "drawn from the letters of my father."

Fantina Polo appears in two Rossi documents. The first map covers Europe, northern Africa, and Asia; it is signed "Fantina Polo," with the year "1329" (see pl. 6). The second is a map of the northeast reaches of

Asia, set in an oval outline (see pl. 7). The map has a brief text accompanying it, which is signed "Fantina Polo," with the year "1329." As was the case with Moreta Polo, this date is again plausible, since records indicate that Fantina died some time between 1375 and 1380.

The Polo Family and Some Notes on the Narrative

Despite the fact that these documents imply a close connection between the three daughters and Marco Polo's travels, nothing in the surviving Polo narrative itself discusses this possibility. Furthermore, Marco Polo's last will and testament does not explicitly mention any maps—maps that might be have been passed on to his daughters.

In fact, considering how frequently Marco Polo and his narrative are discussed today, it is worth noting how little we actually know about the man and his family. As one author succinctly puts it: "La genealogia della famiglia Polo non è troppo chiara nè sicura."[6] Details such as the exact meaning and origin of the nickname often connected to Marco Polo and his narrative—*Il Milione*—are also uncertain.[7] One scholar argues that the name was simply applied to Marco Polo in error and actually referred to another member of the family.[8] In error, too, Ramusio explained the nickname *Il Milione* "as coming from the supposed revenues of Marco Polo, which amounted to ten or fifteen million of a gold denomination," and a number of later historians apparently "accepted this explanation."[9] But the term does not make sense in relation to Marco Polo, and indeed, as the historian John Larner notes, "Great riches are not at all in evidence. Marco Polo ended his days in modest patrician style."[10]

The surname Polo does not help us navigate through this history, because other—apparently unrelated—families of the period shared this name. Marco Polo, the famed traveler, belonged to the Polo family of San Giovanni Crisostomo (or Grisostomo). This was the name of a neighborhood of Venice; one can still visit, in fact, the church of San Giovanni Crisostomo, and behind it La Corte del Milion. A plaque there reads, in part: "Qui furono le case di Marco Polo che viaggiò le più lontane regioni dell'Asia e le descrisse."[11] Another Polo family hailed from San Geremia, located in another district of Venice, and included a figure named Marco Polo.[12] But the two families do not seem to have been related.

The family of our Marco Polo, the famed traveler, has been traced across a few generations, but we have little after the fourteenth century (see app. 3). The record begins with Andrea Polo; he had several sons: Maffeo, Nicolo, and Marco. Nicolo was the father of the famed Marco Polo (*il viaggiatore*), while Nicolo's brother Marco (sometimes labeled *il vecchio*, to avoid confusion) was, then, the uncle. The family was a merchant clan—in this period Venice carried out extensive overseas trade, with connections in Alexandria, Constantinople, the Black Sea, and beyond.

The dates we have for Marco Polo are 1254 to 1324, but they are less clear for his wife, Donata. As far as we know, the couple had three daughters: Fantina, the eldest whose birth date is unknown, but who lived the longest of the three Polo sisters and was still alive in 1375; Bellela, who died in 1326; and Moreta (or "Moretta"), the youngest daughter, whose birth date is also unknown but who we know was still alive in 1348. Besides this information, we have virtually no genealogical data concerning the family. In fact, it was not until over a century after the time of Marco Polo that we find someone writing about him. As one author notes: "The first man who attempted to put together the facts of Marco Polo's personal history was his countryman Gian [i.e., Giovanni] Battista Ramusio (1485–1557). His essay abounds in errors, but prepared as it was at a time when traditions of Marco Polo were still rife in Venice, it does contribute an essential element to any full study of the subject."[13] I will examine some aspects of Ramusio's treatment of Marco Polo's narrative later in this book. But I can note here that nothing in Ramusio explicitly connects Marco Polo and his daughters to maps in the way that the Rossi documents do.

In terms of actual travels, it seems that Marco, his father Nicolo, and his uncle Marco *il vecchio* were the only family members to venture east. The family, moreover, does not seem to be mentioned in Chinese sources.[14] This in and of itself may not be that strange, as other earlier adventurers to Asia often also do not appear in Chinese accounts. Moreover, the Polos affiliated themselves with the "counsels of the Great Khan," who was, of course, a Mongol and not a Han Chinese. Many scholars have agreed, in fact, that Marco Polo never learned Chinese, and this would also imply that he never had a Chinese name, which virtually all longtime residents in Chinese culture are given, even today.[15] The lack of a Chinese name might explain why, then, Marco Polo is not readily found in Chinese sources.

Marco Polo himself was still a teenager when he set out on his travels, though his father and uncle Maffeo had already traveled extensively for trade with the Mongol Khanate.[16] What we might suppose we know of Marco as a person, such as his diligence in serving the Khan, comes primarily from the narrative of his travels, as well as the description in Ramusio. Despite the paucity of information concerning Marco Polo's life, we know that his narrative—even during his lifetime—was translated and published in a number of different languages.[17] This has led, in part, to the very complex history of the narrative manuscripts. One of the peculiarities—to a modern reader of Marco Polo's travels—is the fact that the work was probably first written in a French dialect, and not Italian, although the dialect was indeed "influenced by Italian" and "strongly marked by Tuscan and Venetian."[18] By comparison, it is worth noting that the maps and other documents in the Rossi Collection are almost all in Italian, with short passages, phrases, and terms in Latin, Arabic, and Chinese. Moreover, the Italian in these materials displays aspects of a Venetian dialect.

The original manuscript of the Polo narrative does not survive, and the work itself, as Larner explains, is "not without a certain opacity."[19] Even the identity and role of Rustichello da Pisa—Marco Polo's fellow prisoner[20]—is not fully clear. The early French and Franco-Italian versions of the narrative open with a passage stating that the work was dictated by Polo to Rustichello during their incarceration together.[21] Larner adds, however, his belief that "Rustichello's text . . . was not simply dictated from Marco's memories, but relies on his notes as well."[22] This implies that Marco Polo brought back notes from his travels, which, of course, would lead one to ask whether he also brought back maps. Moreover, one would expect, then, that the narratives deriving from this initial Rustichello text might include mention of the "Marco Polo Maps" and related materials that make up the Rossi Collection. But, as we shall see later in this book, the connections between these documents and the extant versions of the narrative are, in fact, quite sparse.

Again, much of what we know about Marco Polo comes from the narrative. Other works that even mention Polo and his travels are very few in number. Best known is the fourteenth-century *Imago Mundi*, by the Dominican friar Jacopo d'Acqui. This chronicle mentions Polo's being taken prisoner and sent to Genoa but does not give us many details that help us understand the traveler himself, his journeys, or what he might have brought back with him in terms of notes, maps, and so on.[23] A fuller description of Marco Polo is found in an edition of the narrative put together by Ramusio. But this was written long after the time of the traveler himself—in fact, in the mid-sixteenth century—and at the same time contains much additional material and various errors.[24]

Despite its reputation, the narrative is "not a story of adventure and it is not a description of travels."[25] Thus, the portrait we are given of Marco Polo himself is not one of an adventurer—or even a traveler—in that the narrative does not allow us to "map" the journey from one place to another very clearly. The narrative suggests a picture of Marco Polo as a merchant, in its extensive focus on "commercial goods."[26] But the narrative does not flesh out such a picture very precisely either. Another commentator has suggested that "Marco Polo was neither a merchant nor a missionary . . . and least of all an adventurer," but goes on to describe him as a "man of the world and man with a mission."[27]

Despite the interpretation of Marco Polo as *not* being an adventurer, certainly the narrative implies a man on a journey that was filled with observations and experiences, even though they appear in the narrative in the form of somewhat repetitive or generalized statements. Given that here we are examining a series of maps, we should look at Marco Polo in terms of a traveler possessing some geographic knowledge. Yet certainly it is very difficult to retrace his journey from the narrative itself, even as a number of writers have tried.[28] While the internal evidence of the maps discussed in our book here suggest a direct connection to Marco Polo as a traveler *and* a mapmaker, the narrative does not give one a picture of someone carefully charting his extensive sojourn. At best, the narrative reveals Marco Polo as having some grasp of "chorographical and anthropological geography."[29] "Chorography" can be defined, roughly, as a physical description of a locale rather than an accurate mapping or quantitative study of a broader region. Anthropological geography is more clearly evident in the narrative, as Marco Polo goes from place to place describing people and customs, even if briefly.

In fact, "It was some time before cartographers are found to be using [the Polo narrative] . . . as a source."[30] Early fourteenth-century cartographic works such as that of Fra Paolino Minorita seem to ignore Polo, in fact. In Paolino's opening treatise on a world map, the *De mapa mundi* (part of his long *Chronologica magna*), we find many sources mentioned—Solinus, Pomponius Mela, Eusebius, Bede, et al., but no Polo. Similarly, Pietro Vesconte, the maker of a number of portolan charts, did not include Polo material on his world map.[31] We have, then, a paradox of sorts: while apparently there were maps drawn directly based on the travels of Marco Polo—as suggested by this collection here—at the same time, those travels seem to have had little influence on mapmaking of the period in which he and his daughters lived. In other words, the Polo narrative was neglected and the "Marco Polo Maps" lay unknown.

The "truth is that the first maps known to us which were strongly influenced by Marco's Book [sic] and which still remain to us are those in the *Catalan Atlas*."[32] The *Catalan Atlas* is a series of maps on vellum, drawn by a Jewish cartographer named Abraham Cresques in 1375. The renderings of regions in Asia include a number of place-names from the Polo narrative. However, even this work appears quite unconnected to the "Marco Polo Maps" examined here.

Returning to Marco Polo himself, unfortunately what he left behind also does not tell us much about him either. The fact that the "golden tablet" given to him by Kublai Khan is mentioned as being one of his remaining possessions adds some credibility to the narrative and to Marco's claims of having traveled so extensively in service to the Mongols.[33] Marco Polo was involved in legal battles late in his life, but perhaps this is not unusual for someone who had such complicated family and commercial ties. Perhaps we can read something of him as a person in that he left in his will a hefty donation to "priests and for pious purposes."[34] Polo also left money and property to his wife and daughters, as well as his servant, who was also released from bondage.[35]

Various documents that mention the Polo family survive. These documents discuss lawsuits, arrangements concerning property, and so on, but we are not given much insight into the Polo family per se, and no mention is made of any maps or related materials.[36] The only vague hint of such a connection we have is a mention of a book—"cum multis figuris," that is, "with many figures"—with the *itineribus* of Marco Polo. This belonged to a man named Marino Faliero (1285–1355), a doge of Venice who was executed for having led a failed coup. After his death, his possessions—several of them connected to Marco Polo—were, unfortunately, dispersed.[37] Perhaps some of the mysterious "Marco Polo Maps" were among these items.

Marco Polo was married to a woman named Donata Badoer, but little is know about her. And what became of their daughters?[38] We do know that his daughter Bellela married a man named Bertuccio Quirini, while Bellela's sister Moreta was married twice—once to Ranuzzo Dolfin, and later to Tomaso Gradonico. The third daughter, Fantina, was married to Marco Bragadin. We have some record of descendants from Fantina and her husband.[39] Again, for a family as renown as the Polos, surprisingly little survives in terms of genealogies, documents, and historical mentions. For now, we can only speculate that the "Marco Polo Maps" began their centuries-long journey to the present owner by way of these daughters' families.

TWO

Who Was "Biaxio Sirdomap"?

The "Sirdomap Map" takes its name from an unknown Syrian mariner, a certain "Biaxio Sirdomap." The map shows northeastern Asia, with toponyms in Arabic lettering; below it is a short text, apparently written by Bellela Polo; it mentions this Syrian and includes the claim that he knew her father, Marco Polo. Before examining the many questions the text raises, let us take a look at the map (see pl. 1).

A Set of Toponyms and a Story

The map has toponyms referred to by the Roman numerals I–V, a characteristic we see in two of the other maps in the collection as well. On the reverse side, this document has an attached, oval-shaped tab on which is written "Marta Veniero da Padova." Bagrow, in his 1948 article on the Rossi maps, reproduces this document and gives a translation of the accompanying text.

The key given in Italian at the end of the text reads as follows:

I / penisola de li zervi // [i.e., "Peninsula of the Stags"]
II / penisola phoca marina // [i.e., "Peninsula of the Marine Seals"]
III / Valle conzonta e giazata // [i.e., "Connected and frozen valley"]
IV / Isola de le femene // [i.e., "Island of Women"]
V / Eolfo [i.e., Golfo] Mangi // [i.e., "Gulf of the Mangi"]

The map appears to depict regions in northeast Asia, and some of these terms, as we will see, have echoes in the actual Polo narrative.

The text is an account apparently by Bellela Polo. Here is Bagrow's translation with some slight modifications:

So that the noble readers may find more delight concerning the Kingdom of Women in China and the Far East, my father Marco Polo wishes me, Bellela, to show this world map which he obtained from the pilot Biaxio Sirdomap who, for at least thirty years, had sailed the coasts of Asia from Syria to the Far East, trading in seal skins. On this world map, one can see more clearly how Master Marco Polo sailed from the Gulf of the *Mangi* to the east as far as the Peninsula of the Stags, where he met the pilot Sirdomap, who then guided him to the Island of Women, situated to the north and west.

And moreover I tell you that according to what Sirdomap wrote, in that region there is a society whose oracles told them that in the distant past, this people, on account of a scarcity of food, abandoned the caves in the *Auzci* mountains, and from Scythia crossed Asia and entered into that region where they now reside.

Each province is named in their mixed Scythian and Tartar language; for that reason, Sirdomap named them also in the Syrian, as one sees written here.

Never could a Latin or Asiatic man reach this long and narrow island without being killed by the arrows [shot by the inhabitants] in defense of their chastity and superhuman beauty.

But Master Marco Polo, because he took with him a maiden, was able to enter the palace of that queen who treated him courteously and gave him precious stones.

This is indeed a very peculiar story, and one that does not appear in the traditional Marco Polo narrative at all. Bagrow provided some analysis in his article and noted the opinion of the famous Orientalist scholar Hans von Mzik (1876–1961) concerning the text. But both Bagrow and Mzik turned up relatively empty handed. The toponyms on the map indeed seem to be in Arabic, or rather perhaps the Arabic alphabet has been used to render words or names from another language.

As part of the present study, Amir Harrak, of the Department of Near and Middle Eastern Civilizations at the University of Toronto, attempted a translation of the Arabic toponyms. He rendered the toponyms, going from west to east across the map, as follows:

[1] the land of China
[2] Ghadhil [?]
[3] Linwan [?] Castle

[4] Kalb al-Bahr [i.e., "Dog of the Sea"]
[5] Jalid [i.e., "frost"]

Two of these seem to match readily with the Italian toponyms:

Kalb al-Bahr [i.e., "Dog of the Sea"] = "II / penisola phoca marina //" [i.e., "Peninsula of the Marine Seals"]
Jalid [i.e., "frost"] = "III / Valle conzonta e giazata //" [i.e., "Connected and frozen valley"]

The other toponyms are less easy to interpret.

Bagrow did not look at the name Biaxio Sirdomap at all; clearly, it is an Italian rendering of a foreign name. The text tells us that this sailor comes from Siria—that is, Syria—and that he speaks Siriaca. Marco Polo's narrative as we know it today contains no mention of any of the details we see in the text here, and it never mentions a character named "Sirdomap." The name itself is peculiar as well, and perhaps is a corruption of an Arabic name, or a mishearing or crude transcription of such a name. Biaxio seems to be from a Venetian dialect of Italian, with the typical substitution of *x* for a soft *c* or *s*. Thus, the Italian name Blasio is sometimes found written as Blaxio. The name as it appears here—Biaxio—is a dialect version of the Italian name Biagiotto.[1] Are we dealing here with a Syrian who has been given a Venetian first name? The last name does not seem to be found in any other historical record.

I found a very vague connection concerning mariners and maps in the traditional Polo account. Even though the narrative seems to lack much precise cartographic detail, in a mention of the "Island of Seilan" (Ceylon), Polo says: "You must know that it has a compass of 2400 miles, but in old times it was greater still, for it then had a circuit of about 3600 miles, as you find in the charts of the mariners of those seas."[2] Although the translation given by Yule here is "charts of the mariners of those seas," one of the original (French) texts has "selonc qe se treuve en la mapemondi des mariner de cel mer."[3] Note that the actual term used is *mapemondi*—that is, "world map"—most definitely not a mariner's chart. Leonardo Olschki, a well-known scholar who wrote on Polo, states that "Marco Polo was able to consult the charts which he calls 'mapemondi,' that were used by sailors in the course of their difficult navigation among so many lands and islands of the Indian Ocean."[4] Another author also says that these *mapemondi* "can only refer to a nautical map of some sort."[5] But it is not clear why the narrative would use this wrong term for nautical charts here; we should, in fact, ponder whether Polo was talking about a world map or a sea chart—or even a regional map, such as the peculiar "Sirdomap Map," examined above (see pl. 1).

In another part of the narrative, we read of "the charts and documents of experienced mariners," and here the original text is "le conpas et la scriture de sajes mariner."[6] In this case, the word *conpas* does indeed more accurately refer to a chart. Olschki here seems to be more on the mark when he suggests that this is a reference to "track charts and navigation manuals used by Arab and Persian sailors."[7] In fact, *conpas* comes from an Arabic term used by the famed navigator and writer Ahmad ibn Majid (born in the first half of the fifteenth century); that term is *al-qunbas*, which is "used . . . in connection with portolan charts."[8]

Trading in the Farthest East

The "Sirdomap Map" is related to another document in the collection, a short text in Arabic script followed by an Italian text, with the year "1267." Allegedly, this text is written by Sirdomap himself (see pl. 2). The document's short passage in Italian claims to describe the Arabic text above it. As for the Arabic portion of this "Sirdomap Text," an attempted translation is as follows: "Happened 300 years ago [?], departed from Syria to the land of frost ———— the skin [?] of the sea lion ———— there, living [?] ———— they speak a local tongue, fit for him/it [?] ———— Tartar [?]."[9] Below that text is the year "1267" written in Arabic.

Bagrow's translation of the Italian text in this document is reasonably accurate; what follows is a slightly revised version:

The above manuscript dated 1267 / The captain Sirdumap [*sic*; in the other documents here, the name is given as Sirdomap] has sought [?] to provide a few words testifying to the celebrated explorer Marco Polo's thirtieth year of navigation through the seas from Syria along the maritime coast of all of Asia to the farthest east, to a peninsula there that he called "marine seal," trading in seal skins. Just as if he himself were speaking, he writes in detail concerning that remote peninsula washed by the seas, where the people, because of the extreme cold, live in caves. That which provides greatest satisfaction is what he has written concerning the mixed Scythian and Tartar language spoken by those people, which makes one believe that they thus must be related as much to one as the other. All this agrees with that which was recounted by Marco Polo, that according to the belief of the local astrologers, in early times the inhabitants of the *Auztci* mountains migrated from Scythia and joined with wandering Tartars, settling in that region where there abounded every species of fish, and thus [they had] food, clothing, and trade. Finally, worthy of praise is he who rediscovered the *Auztcu* people, who gave to us the use of the precious fur, and guided our Venetian to the most remote corner of the terrestial globe.

In this "Sirdomap Text," as in the "Sirdomap Map," we read of the Auzci or Auztci mountains and the Auztcu people. Because of the rendering of the name (again, perhaps in a Venetian dialect of Italian?), initially it seems difficult to find any connection to existing historical sources. Is the Italian an attempted rendering of the Tartar or "Scythian" language? But we find that the term is very likely the same as Ptolemy's Auxacia, a series of mountains in China—a peculiar connection between Ptolemy and Polo, which we find in other materials in this investigation as well.[10]

We also see here the mention of place "where the people, because of the extreme cold, live in caves." The original Italian text reads: "ove el popolo dal freddo extremo che existe, vive nelle caverne." Mysterious as this sounds, an echo of the phrase is found in some descriptions on the Fra Mauro map, which was made in the middle of the fifteenth century and used Marco Polo's narrative as one of its sources.[11] In several mentions of an ancient northern people known as the Permians, Fra Mauro's map has the following:

These Permians are the last people to the north of the inhabited world . . . They live on wild game and wear animal hides; they are men of bestial habits, and to the very north they live in caves and underground because of the cold . . .[12]

These Permians who live more to the north make their houses underground because of the winter's great cold.[13]

Even though the Fra Mauro map locates these people in the regions northeast of Europe and not Asia, note that here, too, we have people who live *in caverne*. However, there seems to be no further connection, and the geographic and historical description—with its discussion of the Scythian and Tartar peoples—given in the "Marco Polo" documents here strikes the reader as unique.

In one of the other documents in the collection, there is another further brief mention of the mysterious Syrian, although not by name. In the "Moreta Polo Map 1" (see pl. 8), we read of a "Syrian sailor in the service of Master Marco Polo," and this figure speaks of a remote northeast Asian peninsula "situated . . . 8 hours from the islands off Mauritania in the Atlantic Ocean." We will examine this "Moreta Polo Map 1" later in the book.

An Account by Bellela Polo?

There is one final document in the collection that speaks explicitly of the mysterious "Sirdomap": the "Bellela Polo Chronicle" (pl. 3). This is a fairly long text in Italian describing the relationship between the Syrian navigator and Marco Polo. The text purports to have been written by Marco Polo's middle daughter, Bellela. Again, Bagrow's translation is reasonably accurate, but some corrections are needed:

[The] dear princesses, duchesses, marchionesses, countesses, and all the ladies whom it would please to hear about the Realm of Women in China and the Far East, will be able to read that which my father recounted, and I, Bellela, have written from that account.

   After Master Polo had become known in all of China for his zeal, the wife of *Fafur*, queen of the women in the province of the *Mangi*, entrusted to him a message for *Fusint*, queen of the women in the Far East, and put him in command of twenty Chinese and Saracen sailors, and with a big ship he set sail from the Gulf of [the] *Manji* and along the chain of small islands that cross the promontory on the east side of that gulf, and then navigated to the east. Then he entered the ocean where suddenly there arose such a terrible storm that the compass needle swung from this side to that, forcing him to sail to the north side of a chain of islands that enclosed the sea, and stretched east as far as a peninsula where Master Marco Polo disembarked from the ship, twenty-eight days after having departed from China. In this land, he was greeted by a Syrian named Sirdomap, a trader in pelts, who displayed great joy at his arrival. He had been trading in pelts for a good thirty years, and many times [he had gone] as far as another peninsula towards the north and east called Marine Seal, which is twice as far from China. There, there are people who speak the Tartar language, and [something] hardly anyone could believe, had they not seen it themselves: the people go about dressed in sealskins, living on fish, and making their houses under the earth. This peninsula adjoins a muddy valley, and there is such a large ice [ice sheet? glacier?], that certainly one who approached would meet an abyss. He [i.e., Sirdomap] offered to Marco Polo a chart of his navigations. In this manner, they spent five days together in great happiness. But one day, while all were at the table, it happened that the two hundred wives of the caliph approached and began singing and dancing. But they were menaced

[?] by the archers, and three Chinese and one Syrian were shot. So Master Marco Polo decided to try his four arquebuses that he had had made in China; the noise of the explosion made the archers flee in fear. But the caliph, shocked by such terrible weapons, fell at the feet of Master Marco Polo, and made a tribute to him in jewels. Then he entrusted him with a mask of gold for the king of the Tartars. But on account of that threat, Master Marco Polo decided to depart from that country and to go to the kingdom of *Fusint*, asking the advice of Sirdomap, who decided to go with him to that realm. So, with both ships they set sails towards the north and west, and in less than twelve days they reached a long and narrow island, and entered a gulf where they did not find a city, but [instead] found a palace completely covered in solid gold. In this villa was *Fusint*, seated on a golden chair, in command of a band of two thousand ladies called *Bikerne*, with lances in their hands, dressed in ermine pelts, adorned with pearls and jewels. So beautiful were those ladies that they seemed like nymphs. When the queen knew that Master Marco Polo had arrived, she received him with great courtesy, and entrusted him with a lance made of gold, adorned with jewels, for the queen of the women in China. Then he left that island and went by sea to the west. In the city of *Quisai*, he found a thousand ladies with their queen; and when this great lady saw Master Marco Polo again, she was delighted, and rewarded him with many jewels. The king and his Tartar barons also made great rejoicing and festivities for him.

May it be a delight for readers [?]. So may it be [*ita fiat*].

Nothing like Bellela's account appears in the traditional narrative of Marco Polo—a mission for a queen, a journey to distant islands, and other adventures. The text gives us a number of details concerning people and places, some vaguely familiar, but most very unfamiliar. We are told that the people of this region speak the Tartar language, and that they use sealskins for clothing, and—perhaps not surprisingly—that they eat fish. Again, see a faint reflection of this, perhaps, in an inscription on the Fra Mauro map, where he says that the Permians "live on wild game and wear animal hides; they are men of bestial habits, and to the very north they live in caves and underground because of the cold." The account in the "Bellela Polo Chronicle" (see pl. 3) seems to reflect an idea that inhabitants of the farthest reaches of Asia crossed over to the extreme northwestern part of North America. In fact, this idea has a history that predates modern anthropological studies in this area, and speculations about such a connection between Asia and America date back at least to the eighteenth century.[14]

We are also told a bit more about the Syrian mariner Sirdomap. The story seems reasonable enough, but the passage concerning a "palace completely covered in solid gold" and the mention of bejeweled women armed with lances remind us of many of the more mythical stories of early exploration. However, the traditional Polo narrative in fact mentions a palace of gold, but in the context of a discussion of Japan.[15]

No other historical documents or texts seem to corroborate this story. Apparently, there are no other accounts of a queen named Fusint, or a group of women known as Bikerne, for example.[16] The document is peculiar, too, in the sense that it does not stand in isolation, but is related in terms of content to the preceding map and texts concerning "Sirdomap." Any hypothetical fabricator would have had to create all these documents, and carefully coordinate their content.

Some Hints from the History of Cartography

The account of voyages by Marco Polo and Sirdomap in the far northeast reaches of Asia is also peculiar, to say the least. But here and there in the story we find odd echoes of other accounts from various sources. For example, in the documents above, we read of a *penisola de li zervi*, that is, a "Peninsula of the Stags." In earlier times, the Chinese said that the people of the farthest reaches of northeast Asia did not possess domesticated oxen, sheep, pigs, and so on, but instead utilized deer (or reindeer). As cited above, we read in the last text about Sirdomap the following description:

He had been trading in pelts for a good thirty years, and many times [he had gone] as far as another peninsula towards the north and east called Marine Seal, which is twice as far from China. There, there are people who speak the Tartar language, and [something] hardly anyone could believe had they not seen it themselves: the people go about dressed in sealskins, living on fish, and making their houses under the earth. This peninsula adjoins a muddy valley, and there is such a large ice [iceberg? glacier?], that certainly one who approached would meet an abyss.

A world map by the famous Jesuit missionary and scholar Matteo Ricci (1552–1610)—a map that used Chinese sources—has an inscription on an unnamed arctic island: "The inhabitants live in holes, clothe themselves in skins, and do not know how to ride. In this region the cold is so excessive that the sea turns into ice. . . . They scoop out holes in the ice and catch large fish in great numbers."[17] Another inscription in this same northern region is "[T]he natives go abroad by night and conceal themselves by day. They flay deer and clothe themselves in the skins."[18] One element in the story in the "Bellela Polo Chronicle" *does* have a connection to the traditional Polo narrative—the mention of the "city of Quisai." Polo describes this city—spelled as Quisai or Quinsai—as being built upon the water, along with canals and bridges. There is also mention of a "thousand ladies" there, but the narrative describes them as being in the king's seraglio.

The texts and map above also make reference both to a "province of the *Mangi*" and a "Gulf of the *Mangi*" (or *Manji*). In the traditional Polo narrative, southern China was referred to as Mangi or Manzi, an appellation we find in other early European sources as well. It is probably a rendering of the Chinese *Manzi* (蠻子), a term used in the past to refer to southern China.

The "Bellela Polo Chronicle" also speaks of "the wife of *Fafur*," and calls her "queen of the women in the province of the *Mangi*," and indeed, in the existing Marco Polo narrative, we find this same term, rendered in the various versions of the Polo text as *facfur*, *fanfur*, and *fafur*. The term is a transliteration of the Persian designation for the Chinese emperor; its literal meaning is "son of God," and it indeed may be a literal translation of the Chinese term for emperor, *tianzi* (天子), which means "son of heaven."[19]

In the same passage in the "Bellela Polo Chronicle," we read that the wife of *Fafur* gave Marco Polo a message to carry to a queen in the Far East. But the traditional Polo narrative says of Fafur only that she was taken to the court of Kublai after the fall of the Sung capital. There is no mention in the narrative of her giving a message and command of a fleet to Marco Polo.

Nor, in fact, do we find any other source that mentions a "*Fusint*, queen of the women in the Far East." We do, however, find in the Polo narrative the "Island of Women"; in fact, that is a term that we see in a great number of other early geographic texts. It is not clear whether the "Bellela Polo Chronicle" text here refers to an "island of women" or simply a kingdom of women. Note that in the famous map of Fra Mauro, there is an inscription that reads, "Loco habitado per femene bellicose e valente e querizano tra esse."[20] A commentator points out that "the authors Fra Mauro generally uses as sources for his description of these regions make no mention of such women warriors," but of course we do indeed find such a description in our Polo maps.[21] Nothing else in Fra Mauro's map seems to match what we see in the "Polo" documents in the collection here.

WHO WAS "BIAXIO SIRDOMAP"?

A number of ancient Chinese legends also discuss islands or kingdoms of women.[22] The ancient Chinese tale of Hui Shen, which we will look at later in this book, not only shares a number of toponyms with the Rossi documents, but also has a passage concerning a kingdom of women. Mysterious as they are, these maps and texts reveal a unique fusion of Western geographic knowledge and Eastern lore.

THREE

To the Distant East

The "Map with Ship" resides in the Geography and Map Division of the Library of Congress (pl. 4). John Hessler, a senior cartographic specialist there, carried out a brief study of this map. He noted that the study comprised two parts. First, there was C-14 dating of the vellum, which "showed two peaks, one 1463–1526, the other 1556–1633."[1] This is interesting, of course, but the problem is that the ink was *not* tested, so no further conclusions can be reached in terms of the dating and history of this document. Next, in terms of studying the actual cartographic rendering, Hessler commented that no particular investigation was done, "except [to] look for palimpsest regions, in which we did not find anything particularly interesting."[2]

A more in-depth summary of Hessler's thoughts on the "Map with Ship" unfortunately has not been published beyond a short, but thoughtful, "blog" entry:

[The] "Map with Ship" has no internal or geometric inconsistencies that would lead us to believe that it was definitively copied from a modern map. However, the models employed in coming to this conclusion unfortunately do not totally rule out this possibility. For although we have suggested that the map was not copied verbatim from a Portolan chart we have no way of narrowing down the possible geographic sources more precisely at the present time. All we can say based on this study is that it is still possible that the Rossi map was copied from or based on geographic sources that are consistent in their construction, geometry and scale error with those that COULD have been produced or copied from late medieval and early modern sources. The question of the date of the map and its authenticity must, however, await further studies.[3]

The "Map with Ship" itself is a strange amalgam indeed: there is the cartographic image itself, a series of peninsulas and islands in a kind of stylized frame. To the left, there is a rather crude rendering of a sailing ship. The map covers the northeastern part of Asia, a strait, and land beyond. The parchment appears as if it has been written on at different times, and there are bits of writing that have been scratched away.

Chinese Characters and Italian Toponymy

In terms of content, the "Map with Ship" is closely related to several other works in the collection. There is a string of crudely written Chinese characters, apparently copied from a Chinese source by someone not familiar with that writing system. In his 1948 study, Bagrow noted that the famed sinologist Bernhard Karlgren could only decipher a few of the Chinese characters, and those could be translated as "their names come from olden times." The characters have been crudely copied, it seems, and indeed only a few can be read with any assurance. From the eleven characters in the inscription, one can make out the following characters with a measured degree of certainty: 丁, 的, 百, 出, 七, 四, 由, and 家.[4] As a whole, however, there seems to be no way to translate these characters into a proper sentence or phrase.

On the map itself there are toponyms in Arabic script, but some areas of the map have a numbered key, with the Roman numbers I through IV. These numerals correspond to a list in Italian that is found to the left of the map, below the drawing of the ship:

I. India e pertin[en]te isole segondo come dixono li saracini. II. Cattigara de tartaria isole de Zipango e isole pertinente. III. Peninsola de li lioni marini. [IV] Isole consonte a la peninsola de li servi situata a IV hore de varietade de le provinzie amurade de tartaria.

[I. India and adjacent islands, according to what the Saracens say. II. Cattigara of Tartary, islands of Japan, and adjacent islands. III. Peninsula of the Sea Lions. IV. Islands connected to the Peninsula of the Stags, situated at four hours difference from the walled provinces of Tartary.]

The term "Saracens" perhaps refers to Muslim navigators, who seem to be suggested as one of the sources for this map. The mention of a "Peninsula of the Sea Lions" and the "Peninsula of the Stags" reminds us of identical terms on the "Sirdomap Map." In the actual Polo narrative, the term *dicono li saracini* is used, in fact—but only in a brief mention concerning Ceylon.

Between the "Peninsula of the Sea Lions" and the "Peninsula of the Stags," the map shows a narrow strait. This strait is very much like one that we see in European maps of the sixteenth century and later. The cartographic figures on the "Map with Ship" bear some similarity to the cartographic concepts of the Italian mapmaker Giacomo Gastaldi (1500–1566), who also had a narrow strait separating Asia and the North American mainland. We find other representations that are also somewhat similar to that of the "Map with Ship"—particularly its depiction of the regions around the strait—in Bolognino Zaltieri's map of 1566, in the 1574 map of Paolo Forlani, and in an eighteenth-century map by Aaron Arrowsmith. These similarities will be discussed later in this book.

Understanding the Remote Northeast

Precise knowledge of the farthest northeastern regions of Asia—and the northwest part of the American continent—took a long time. For decades, it was uncertain whether Asia and America were connected or separated by water, and if the latter, by how far. In this region, some mapmakers "left the region largely blank, without committing themselves to a theory."[5] In the period from 1540 to 1600, cartographers tended to put a strait between the two landmasses, although this depiction was not—as far as we know—based on any actual exploration of the area.[6] An expedition by the Dutch navigator Martin Gerritsen de Vries in 1643 actually complicated the matter. He sailed as far as Sakhalin but believed that he had found the waters separating Asia and America. This conception influenced another series of maps, which included peculiar lands such as Yezo (also written as Jesso, Iesso, etc.), "Compagnies Landt," and so on in these distant northern reaches.[7] One such example is the depiction of the northeast portions of Asia by the German cartographer Johann Baptist Homann (1664–1724).[8] Interestingly, a map created by Matteo Ricci—who resided in China and used Asian sources—has the following inscription: "Strait of Anian. . . . At this point it was formerly thought that the lands of the two sides of the world were joined together. But now it has been discovered that this great sea intervenes, and that through it one can penetrate to the Northern Ocean."[9] Unfortunately, we have no record of such discoveries at this early date, and it is not clear where Ricci obtained his information on Anian here. Later, there were Russian expeditions that reached the Aleutians, and these provided some clarification of this area.[10] None of these navigations and depictions, however, seems to correlate directly with those found in our "Marco Polo Maps."

It was only with the voyages of the Danish navigator Vitus Bering in 1728 and 1741 that it became clear that there was a strait separating the two continents.[11] As noted above, prior to that, there had been speculations that there was some kind of waterway—the "Straits of Anian"—but others had believed that terra firma connected Asia and America.

The Rossi maps discussed here appear to show input from Asian sources, as they have Chinese writing. This raises the question of whether traditional European maps of these regions ever showed such influences. One example, in fact, does—although it is a fairly late work. The English cartographer Thomas Jefferys produced a map in 1768 entitled "Carte générale des découvertes de l'Amiral de Fonte." This map showed the northern regions of Asia and America, combining accurate mapping of northeastern Russia with somewhat fanciful details of a northwest passage in Canada.[12] But the important point here is that Jefferys placed clear labels on his map stating "partie copié de la Carte Japonoise" and, on a large stretch of land in northeast North America, "Indiqué par les Japonois." Further east, the map also has "Partie Nord est de la Mer de Tartarie representeé dans la carte Japonoise."

Some decades earlier, however, we have also have indications of the use of Chinese sources in the mapping of these remote regions. The Dutch traveler and cartographer Nicolaas Witsen (1641–1717) journeyed to Russia and made maps during his time there. Although stationed in Moscow, in 1690 Witsen wrote to the president of the Royal Society in London that he had "not ceased . . . to send Letters unto and received Answers from the most Northern and North-East parts of the world."[13] This research led to Witsen's coming into possession of maps of Siberia from the Russians, but also "maps made by Chinese and Jesuits." He obtained, too, "Chinese maps which indicated to the North of Korea various islands and a passage."[14] There were also Chinese sources for "East Siberia up to the coast of the Ice Sea," and—curiously—a "little Arabic map."[15] Witsen also cited a "Spanish map, drawn by a Spanish pilot who sailed from China to New Spain"; on this map, Witsen noted, "to the east of the Vries Strait and Compagnie's Land, there was faintly drawn the coast of a continent, and on its coast, a great many islands."[16] Although Witsen was working in the seventeenth century, one wonders if there is some connection here to the depictions in the "Marco Polo Maps." Especially intriguing, perhaps, is that mention of the "little Arabic map."

Still, the depictions of the "Polo" maps must be examined with care, since they represent a very unusual series of claims: (a) that Marco Polo traveled in the seas beyond China; (b) that he received information from a Syrian navigator there; (c) that he also incorporated Chinese knowledge of these regions, as evidenced by the Chinese text on the maps; and (d) that well before the voyages of Bering, these areas were explored and mapped by Marco Polo, and that this information was passed on to his daughters. Witsen seems to have accomplished some of these endeavors, but that was some four centuries after the time of Marco Polo.

Text from a Golden Tablet

A second map dealing with these little-known regions is the "Pantect Map" (pl. 5). This map provides a broader picture of Asia and approaches the region in some ways that are different from the "Map with Ship." The map takes its name from an inscription on the right-hand margin that reads "Pantect De Praefectoria Potestate Interpret / in lat / Domus de Sanseverinus Urvinum." The term *Pantect* is unclear, but it may simply be an alternate spelling for *pandect*, which is a term for a body of laws or a legal code. In this inscription we see a reference to the Sanseverino family, along with Urvinum—that is, the Italian town of Urbino, a locale with which the family was intimately connected.

Another inscription, in Italian, runs along the left side of the map. This brief text states that an accompanying Latin passage above it is a translation from the original Tartar of the famous "golden tablet." This tablet apparently was given to Marco Polo by the great Kublai Khan. Such tablets are mentioned at a few points in Marco Polo's actual narrative.[17] The Italian text on the "Pantect Map" says:

Interpretation from the Tartar into Latin of the tablet
of gold, for authority and possession of the province and
adjacent peninsula, islands, Southern and
Eastern Oceans, and with bordering subjects and foreign peoples, that

Kublai, Emperor of Tartary, [had] made for Marco
Polo of Venice, and from there [?] the explorer
was rewarded with much treasure.
Done [i.e., written] by Rugerius Sanseverinus.[18]

The Latin that appears on the map—the supposed "interpretation from
the Tartar"—presents a number of grammatical challenges:

Marcus Polu de Venetia patrono ratiorum summarum
adecto inter praet judicio Klubai Kan legato
provincia janiu quae Tatari obediunt /
aliae ab initio per amititiam tibani Tartaris adjuncta //
Multas existimatis insulas in longssimo Oriente
non invisa solum sed etiam inaudita primus
Tartariae in locum entravit anno priore tu
meruisti // tua voluntate parent peregrina
commercia cum conscendis gemmatum tribunal
sed tot testes pateris quot te agimina
circunere cognosce

Ego Klubai Kan super me deo est //

Indeed, the Latin requires some emendation:

Marcus Poli de Venetia patron<us> ratio<n>um summarum
ad<l>ecto inter praet<orios> judicio Kublai K<h>an legat<us est>
provincia<e> Janju <cui> Ta<r>tari ob<o>ediunt,
<et ad> alia<ae> ab initio per ami<c>itiam Tartaris adjuncta<e>.
Multas existimatis insulas in longissimo Oriente
non invisa<s> solum sed etiam inaudita<s>. Primus
Tartariae in locum <i>ntravit anno priore. Tu
meruisti. Tua voluntate parent peregrina
commercia cum conscendis gemmatum tribunal,
sed tot testes pateris quot te agmina
circum<i>re cognosce.

Ego Kub<l>ai <Kh>an <sum>, super me de<us> est.

We may roughly translate the Latin rendering of the "golden tablet"
text as follows:

Marco Polo of Venice, a defender of the highest counsels,
granted praetorian rank by the will of Kublai Khan, legate of the
province of Janju[19] which the Tartars obey
<and of> others joined to the Tartars through friendship from the beginning.
You think that there are many islands in the most distant East,
not only unseen but unheard of. He first
entered the territory of Tartary a year ago. You
have served <me well>. With your <good> will let them do business as itinerant
merchants when you ascend the bejeweled tribunal,
but you permit as many witnesses as the number of regiments
surrounding you, understand.

I am Kublai Khan, above me is God <alone>.[20]

One peculiarity of this Latin text is that one of the lines seems to have
been taken directly from a work by the Roman writer and statesman
Cassiodorus (ca. AD 490–585). In one of his many official letters, we
find the line "considis geniatum tribunal: sed tot testes pateris quot
te agmina circumdare cognoscis," which is almost identical to a line
in the text above: "cum conscendis gemmatum tribunal sed tot testes
pateris quot te agimina circunere cognosce."[21]

It seems that in some way the writer of the Latin here on the "Pantect Map" has carried out a kind of "pastiche" of some kind, using the
Cassiodorus text to supply a concluding formulaic line to the apparent "golden tablet" text. We find similar bits of this elsewhere here.
For example, the Latin text begins with Marco Polo's name, and then
"patrono ratiorum summarum adecto inter prae judicio . . . legato provincia" which we can emend to "patron<us> ratio<n>um summarum
ad<l>ecto inter praet<orios> judicio . . . legatus provincia<e>." There is
an existing Latin phrase, in fact, that runs "patronus rationum summarum adlectus inter consulares iudicio . . . legatus," with the only difference being that our text has "praet<orios>" instead of "consulares."[22]
This existing Latin phrase, oddly, is found in an inscription rather than
a text. How our author could have come to use it, then, is rather puzzling, unless this text is from a very late date indeed, when the inscription appeared in a published work, for example, Ettore de Ruggiero's
Dizionario epigrafico di antichità romane.[23]

So, again, it seems that the writer of our text here is creating a kind
of "pastiche" or composite, rather than actually rendering Tartar into
Latin. The use of such formulaic Latin passages from other texts is not

unheard of; there was a tradition in the Middle Ages of what is called *ars dictaminis* or *dictamen*, the art of composition and style in writing, especially letter writing. In addition, there were manuals in the Middle Ages which contained standard common forms used in letters and similar documents, with some of these forms coming from early Latin authors. Cassiodorus, in fact, was one of these sources.[24] In short, our writer here may have been using such *dictamen* materials. However, other parts of this Latin "golden tablet" text, despite—or perhaps because of—their awkwardness, may be original in some way.

The traditional Polo narrative and other sources tell us some interesting facts about these tablets. Apparently, members of the Polo family even brought back some of the tablets that had been given to them.[25] In book 2, chapter 7, of the Polo narrative, we are given the text of a tablet: "By the strength of the great God, and of the great grace which He hath accorded to our Emperor, may the name of the Kaan [*sic*] be blessed; and let all such as will not obey him be slain and be destroyed."[26] The golden tablets of the Mongols that actually have been excavated state: "By the strength of the eternal heaven! May the name of the Khan be holy!"[27] Note that this does not quite match the "golden tablet" text on the "Pantect Map." In book 2, chapter 8, of the Polo narrative, we also do not find a match, but we read a brief description of a tablet: "When the Prince [i.e., the Great Khan] had charged them with all his commission, he caused to be given them a Tablet of Gold, on which was inscribed that the three Ambassadors should be supplied with everything needful in all the countries through which they should pass—with horses, with escorts, and, in short, with whatever the should require."[28] That these kinds of tablets were used in this period is a well-established fact, and as just noted, evidence indicates that the Polo family indeed brought one back to Italy.[29] In the will of Maffeo Polo—Marco's uncle—there is mention of a gold tablet from the Great Khan.[30] What that particular tablet said we do not know, and there is no way of determining whether the text given here is from the tablet brought back to Italy. It is possible that the author of our text here genuinely had the Tartar or perhaps even a Chinese text in front of him and was trying to render it into Latin. It is also possible that the author was simply making up what he thought such a "golden tablet" text would say, and so put together a series of "stock" Latin phrases and formulae.

Ptolemaic Connections

Also in the "Pantect Map," below the cartographic rendering, we find another brief inscription in Latin, this time discussing Tartary (see pl. 5): "According to Polo, Tartary is one hundred and fifteen degrees seven hours distant from the Fortunate Islands, toward the East. Venice, 5 July 1297."[31] The "Fortunate Islands" are quite common in early geographic texts and maps and are often interpreted as representing the Canary Islands. In Ptolemy, they are the point farthest west, the point from where measurement of longitude begins. Thus, it is not surprising to find them here used as a starting point for the calculation of longitude. However, the rest of the inscription on the "Pantect Map" is rather ambiguous. It speaks of a distance of "one hundred and fifteen degrees seven hours"—is this meant to be read as "one hundred and fifteen degrees [or] seven hours" or as "one hundred and fifteen degrees [and] seven hours"? In the first case, one should note that seven hours would be equivalent to 105 degrees (each hour being equivalent to fifteen degrees), clearly differing from the figure of 115 degrees. In the second reading, one would have 115 degrees *plus* 105 degrees, for a total of 220 degrees as the distance from the "Fortunate Islands" to Tartary.

The "Pantect Map" shows Asia from the Persian Gulf to the seas around Japan, and there are a number of place-names. Moving from west to east, we see Persia, Arabia, Aethiopia, and India. Then we encounter Serica, that is, China. There we find the city of Cattigara, from Ptolemy, and Campalu—the city of Kanbalu, that is, Beijing, discussed in the traditional Marco Polo narrative.

To the south is the toponym Cangem; this refers to the Ganges River, or more exactly what was called by the early cartographers India extra Gangem, meaning the part of Asia beyond the Ganges. Beyond that area, further to the south, is the Oceanus Indicus. To the east is the Magnus sinus, the "Great Gulf," a toponym also taken from the ancient geography of Ptolemy. Moving northward, we see Cipangu, the term for Japan found in the Polo narrative.[32] So far, so good—all of these place-names are familiar from both Ptolemy and the traditional Polo narrative. But in the far north of the map, we see a string of islands connecting Asia with a land to the east. A similar configuration is in the "Map with Ship," as well as other maps in the Rossi Collection. In this region of the "Pantect Map," we find the label "Transerica pons"—literally, "trans-China bridge," a very curious toponym indeed.

To the right of the map, we have another brief inscription, with a key to numbered islands appearing on the map:

There are many islands, the primary ones being
I / Aualitis or menuthius
II / Salice
III / Agathadaemonis
IV / Jaba diu
V / Satirorum[33]

These place-names are all found in Ptolemy; indeed, a number of the maps in this collection utilize the famous geographic work of Ptolemy, and they seem to represent a very early use of his text. Ptolemy, in his *Geography*, provides instructions on how to create a map of the inhabited world, termed the *oikoumene* in Greek. His text actually includes a list of longitude and latitude coordinates from which one can construct a map. When his work was rediscovered in the Middle Ages, many maps were assembled from these coordinates.

The problem is that the maps in the Rossi Collection are apparently from a very early period—the late thirteenth century—a period before Ptolemy's work had been rediscovered in Italy. In fact, the Byzantine monk Maximos Planudes (1260–1310) found a manuscript copy of the *Geography* around 1295, a time just before the purported date of this map.[34] The work was not translated into Latin until the fifteenth century.

The Rossi maps reflect some knowledge of the *Geography*, but through what connection remains unclear. It is interesting to note that there is actually a quite early *false* connection made between Marco Polo and the geography of Ptolemy. One manuscript of Marco Polo's narrative, a translation in German dating from the latter half of the fifteenth century, "goes so far as to end by making Marco recommend Ptolemy for 'further reading'"![35] As one commentator says, "Here Marco is made to accept the mistaken but common medieval identification of the cosmographer with King Ptolemy Philadelphus II [285–247 BCE]: 'No more will I speak to you of foreign lands. But who wishes to know more will find it in the great scholar and king of Egypt who has not alone written for you of the world but also of the heavens with its stars and the whole firmament. Deo gracias.'"[36] The last line, of course, refers to Ptolemy's other famous work, the *Almagest*. Despite what this German manuscript says—and despite the evidence of the Rossi documents—we have no historical evidence that Marco Polo had access to anything having to do with Ptolemy's works.

A Missing Text

In 1934, Marcian Rossi sent another document to the Library of Congress for examination, a manuscript he described as a "keynote found fastened to the Pandect [*sic*]." The "Keynote to Pantect Map" itself is now missing, surviving only in a photographic reproduction in Leo Bagrow's article (see fig. 3).

The reproduction, however, is sufficient to allow a deciphering of the text. There are two passages, one on the recto and one on the verso, and the descriptions that they provide are connected to another document in the Rossi Collection, the "Lorenzo Polo Chronicle," which will be examined later. The recto text of this "keynote" reads:

Depiction of India and Tartary by Marco Polo, and of the many islands explored by him, so that the Great Khan honored him by giving him authority over a province of his realm. No one had ever navigated toward the east, a desert of sand three thousand miles from the Tartar realm. But Marco Polo, with ten ships, set sail and went by sea so far that he reached a chain of islands and [finally] a great peninsula. There they found caves here and there. They [i.e., the inhabitants] wear trousers and shirts of the skin of seals and deer.

We have seen the discussion of seals and deer in the documents concerning "Sirdomap," and in the "Map with Ship" there is the "Peninsula of the Sea Lions" and the "Islands connected to the Peninsula of the Stags, situated at four hours difference from the walled provinces of Tartary." But what is especially peculiar here is the description of the voyage itself, a trip that is found nowhere in the traditional Polo narrative.

On the verso side of the "keynote" document is a text that seems to be describing the content of the "Pantect Map" (pl. 5): "Universal depiction of Asia, rendered by Marco Polo. According to a hypothesis, there are seven hours or 115 meridian degrees from the islands of Viteperia or Purpuraria of the Fortunate Islands to Cattigara in Tartary. In his manuscript in our language, one reads that on these islands, every man shoots arrows at strangers [the text is not clear at this point] . . . and it [i.e., this region] seems to him to be a place wild and without any use." The discussion in this text is connected closely with that found in other Rossi documents. First of all, we have again—as we did on the "Pantect Map"—the mention of "seven hours" and "115 meridian degrees." Here they are described as being the same, even though seven hours actually would be equivalent to 105 degrees.

In the "Moreta Polo Maps," we find the Fortunate Islands again, notably Junonia, an island discussed in the famous *Natural History* of the Roman writer Pliny the Elder. Here in the "Keynote to Pantect Map," we have Viteperia and Purpuraria; the name Purpuraria is probably a reference to Pliny's Purpuraii, the "Purple Islands," so called because of the purple dye obtained from there in ancient times. Pliny, however, describes these "Purple Islands" as distinct from the Fortunate Isles.[37] Viteperia, meanwhile, does not match any island in Pliny's discussion.

Land beyond Asia

Of course, what is most remarkable about the maps discussed in this chapter is that they portray land beyond the farthest northeastern regions of Asia. These depictions would suggest knowledge of the western shores of North America in the thirteenth century, for which there seems little other evidence. It is interesting to note, however, that there exists a series of later maps that hint at early Chinese knowledge of the northern Pacific Ocean and its boundaries.

The eastern portion of the 1517 map of Pedro Reinel, a Portuguese cartographer, depicts many islands in the Pacific Ocean (fig. 5).[38] One is a small, rectangular isle that appears right near the top of a long coastline on the other side of the ocean; it is labeled "Chis," that is, "Chinese."[39] This map—with its depiction of an eastern coast of Asia, an ocean with a narrowing at the top (where the aforementioned "island of Chinese" is found), and a coastline beyond—seems to suggest some vague knowledge of the Pacific coast of North America.

The historian Albert Kammerer seemed uncertain: in his study of this work, he puts forward no conjectures concerning what he terms the "deux continents" on either side of the eastern ocean. However, he does suggest that the large isle, despite its label "Chis," could represent Japan: "The location of this island corresponds (if you will) to Japan and not to China."[40] Of course, this is conjecture, and one would have to ask, moreover, why the mapmaker would apply the label of "Chinese" to a small island in the distant ocean.

The French scholar E. T. Hamy, in his analysis, said of this coastline to the east: "The map of Reinel ends toward the east with a long, simple line, which must represent, quite by chance, the western coast of the New Continent."[41] The historian Henry Harrisse wrote about a map that seems to be the same as this Reinel map, which he says he saw in the archives of the Staff of the Bavarian army in Munich. Har-

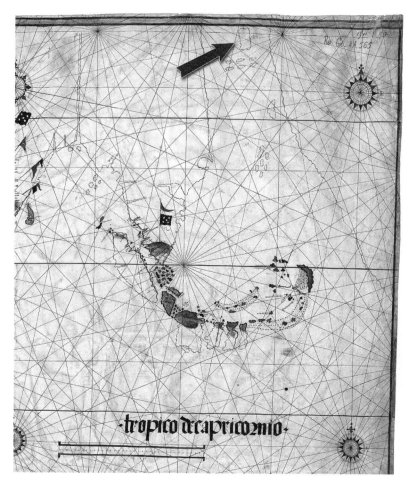

5   The 1517 map of Pedro Reinel; note the small, rectangular island at the very top of the map (marked with arrow). (Source: Bibliothèque nationale de France.)

risse says that the map represented "the Indian seas, and, thus far for the first time, the Malucca islands, which were discovered by the Portuguese in 1511. Its importance to us lies in the unexpected fact that east of the Asiatic coast, there is a continental land running from 4° south to 40° north latitude. What can that continental region be if it is not the New World? And if so, how curious it is to see the Pacific shores so clearly and so extensively depicted at such an early date."[42] The 1522 map of Nuño Garcia de Toreno (fig. 6) has elements similar to the 1517 map of Reinel. We find a large eastern ocean, perhaps derived from the

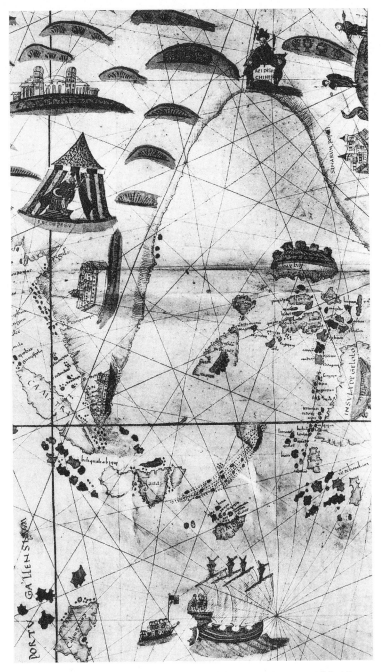

6 The 1522 map of Nuño Garcia de Toreno, eastern portion. (Source: Bibliothèque nationale de France.)

sinus magnus ("great gulf") of Ptolemy. Ptolemy's depiction of the world ended in the east with a large body of water, contained by a coastline that ran around and back down to the southwest (see figs. 7a and 7b). This depiction was used by mapmakers throughout the Middle Ages and the Renaissance. There were some variations and adaptations in the work of later mapmakers, as the Indian Ocean and Southeast Asia became better understood through European explorations.

However, in the 1522 map—a relatively early work in terms of the history of these Pacific explorations—the configuration of the eastern ocean is already significantly different from that found in Ptolemy; here, the coastline curves to the southeast, as in the 1517 Reinel map. The 1522 Nuño Garcia de Toreno map also includes, to the south, islands in the eastern seas recently discovered by the Portuguese, for example, Timor.[43] The coast all the way to the east bears the toponyms Rei de los Chines and Sinarum populi and again goes toward the southeast instead of curving back around to the southwest as in Ptolemy.[44]

Kammerer noted: "One knows of no original chart or world map that encompasses the Far East between 1492 and 1502. On the latter date the famous anonymous Portuguese world map known as the 'Cantino' was drawn."[45] The famous Cantino map was drawn after Vasco da Gama's voyage in 1499 to Calicut, and the beginnings of Portuguese exploration in Asian waters. It is an anonymous and undated Portuguese work, bought by Alberto Cantino in Lisbon in 1502, apparently on a secret mission for Ercole d'Este, the Duke of Ferrara.[46] On the Cantino map, the coast of Asia goes up in a northeasterly direction, and the depiction of the Asian regions in general shows certain Arab influences.[47] This may be an echo of the apparent Arab sources for one of the Rossi maps, the "Sirdomap Map."

The 1513 sketch charts of Francisco Rodrigues represent the earliest surviving cartographic work with depictions of the southeast regions of Asia directly based on Portuguese seafaring knowledge (see fig. 8).[48] The Portuguese, indeed, were the first Europeans to have such knowledge. One of the sketches is also noteworthy in that it is the first to use the term Jampon, rather than Marco Polo's designation of the country, Zipangu, or variations thereof.[49] The Portuguese did not reach Japan itself until around 1543.[50]

The particular sketch shown here presents a cartographic image of a section of some lands in the southeast part of Asia, but is rather difficult to decipher (see fig. 8). The coast is interrupted by two unnamed rivers. There is also an ambiguous inscription, "ate qui tem descoberto os chims," which could equally be rendered as "It is up to here

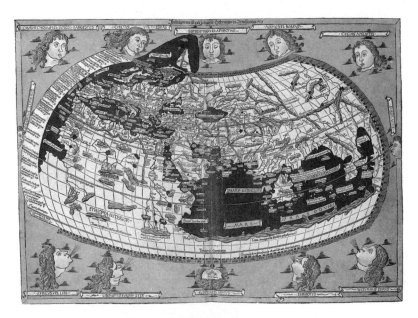

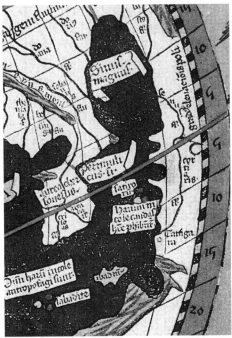

7a A printed version of the world map of Ptolemy, 1482 (Johannes Schnitzer, engraver). (Source: Roderick Barron, *Decorative Maps* [London: Bracken Books, 1989].)

7b A close-up of the eastern regions of Ptolemy, showing the *sinus magnus* at the top.

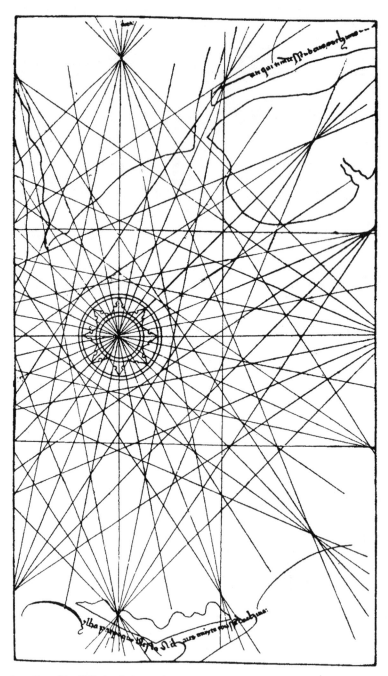

8   One of the 1513 sketch charts of Francisco Rodrigues. (Source, Albert Kammerer, *La découverte de la Chine par les Portugais au XVIème siècle et la cartographie des portulans* [Leiden: E. J. Brill, 1944], pl. XXII.)

that one has found the Chinese" or "It is up to here that the Chinese have discovered."[51] To the south, there is an island with the inscription "Ylha parpoquo. Nesta achão se muytas coussas da China," that is, "The isle of Parpoquo; on this [island], many things [i.e., merchandise] of China are found." Kammerer suggested that one could identify this island with the regions near Korea, but areas so far to the northeast had not yet been reached by the Portuguese in this period.[52]

Notable, too, is the coastline on the eastern side of the seas in these maps—that peculiar distant coast on the other side of the Pacific Ocean. Balboa had reached the Pacific coasts of Central America four years before, but no one had explored the shores of North America far to the north; therefore, it is unclear how this line might represent any knowledge of the western coast of this region despite the apparent similarity.[53]

In his study, Kammerer implied that the coasts on the Reinel map should be seen simply as Ptolemy's *sinus magnus*, with no northern curve at the top.[54] The Portuguese historian Armando Cortesão expressed a similar belief, saying that the eastern coast goes, as we see, to the south, "formando nitidamente o *sinus magnus* da concepção ptolomaica"—that is, "clearly forming the *sinus magnus* ["great gulf"] from the Ptolemaic model."[55] However, a look at the map itself shows that this coast does not finally turn to the southwest as it does in the Ptolemaic system, but rather to the *southeast*.

Furthermore, the Reinel map lacks all the features and the toponyms of Ptolemy's coastal areas in the east. It is also not clear why the mapmaker has left out the curve at the top if he is, in fact, simply recreating Ptolemy's *sinus magnus*. It may be that the depiction of this area is not based on Ptolemy at all, but on some other unknown source, perhaps even an Asian chart or map. These peculiar Portuguese maps, with their strange coastlines, may be the closest parallels we have to the Rossi maps and their depictions of land beyond the eastern coasts of Asia. But a precise connection—if any—remains elusive.

FOUR

The Daughters' Maps

Historical records tell us very little about Marco Polo's daughters, so it is rather suprising that a whole series of documents in the Rossi Collection contain their names and what seem to be comments about their father's travels. The names of the three daughters—Fantina, Bellela, and Moreta—appear in a number of these maps and texts, often with the implication that some of these works were, in fact, penned by them. The "Fantina Polo Map 1" is a map covering Europe, North Africa, and Asia, with the entire depiction covered by a "longitude-latitude" grid (pl. 6). The place-names here are referred to by a series of Roman numerals. The map is signed "Fantina Polo," with the year "1329." Similar gridded configurations can be seen in the "Moreta Polo Map 1" and the "Moreta Polo Map 2" (see pl. 8, and figs. 4a and 4b).

Chinese Place-Names

In the "Fantina Polo Map 1," the place-names referred to by Roman numerals run as follows:

I. Canaria [i.e., Gran Canaria, one of the Canary Islands]
II. India [i.e., "India"]
III. Tartaria [i.e., "Tartary"]
IV. Zipangu [i.e., the early Italian name for Japan]
V. Uan Scian [meaning unknown]
VI. To Qiú [meaning unknown]
VII. Fusan [meaning unknown]

This last term, Fusan, is apparently the same as similarly spelled top-onyms in other maps of this collection, with a possible connection to the early Chinese legend of Fusang.

On the right margin of the map there is a text in Italian:

En lo mare atlantico de Canarie, ante de Africa, plu en l'otra parte de Tartaria ò Serica, enfin che dentra en la grande tera incrosaa nomaa Focaa si è una distancia plu de VIII ore, quand en Europa è sera, alì è doman.

Fantina Polo

Venegia, MCCCXXIX[1]

The meaning in some places is a bit ambiguous; it is not clear what *in-crossa* means here; perhaps it can be read as *incrociato*, that is, "crossed" or "traversed," or even *incrostata*, meaning "caked," for example, with ice. A rough rendering of the above text in English is

From the Canaries in the Atlantic Ocean off Africa, [to] beyond the other part of Tar-tary or Serica, [and] finally [to the] interior of that great land that has been crossed, named Focaa, is a distance of more than eight hours; when in Europe it is evening, there it is tomorrow.

Fantina Polo

Venice / 1329

In this text and in the numbered key, it appears that at some point in the document's history *Fusan* was written over to render it as *Focaa*. On the verso of this "Fantina Polo Map 1," we find the words "Tabula Geogra-phica Fascio 96. Folio 255," an inscription that will be examined below.

More Chinese Place-Names

The "Fantina Polo Map 2" is a map depicting East Asia, a strait, and a peninsula with a chain of islands; the drawing is set in an oval frame (see pl. 7). A series of Roman numerals refer to a set of toponyms. The work is signed "Fantina Polo" with the year "1329." The "Lorenzo Polo Chronicle," which will be discussed later in this book, includes a vari-ant of the text that appears here.

The toponyms below the map are very similar to those found on the first Fantina Polo map, just described:

I / Tartaria ó Serica [i.e., "Tartary or Serica"]

II / Zipangu [Japan]

III / Uan Scian [meaning unknown]

IV / To Qiú [meaning unknown]

V / Ta Can [meaning unknown; note that this toponym does not appear on the first "Fantina Polo" map]

VI / Fusan [meaning unknown]

VII / Maro Oriente [i.e., "Eastern Sea"].

Just below the map, there is a text in Italian:

De quanta città è en Tartaria de saxi tuta mura[t]a d'entorno, tant è pleno lo mare Oriente de Isole cun dua longa penisula inter li quali corrono aque entorbala[t]e en fin en la granda penisula ke li nauti Tartari nomean Ta Can, la prima de nome Uan Scian possa (?) la catena de isole To Qiú, la proxima penisola Ta Can che è conzonta a . . . , tera denuda, alì è l'homo cun spada de avori et zascaduna femena sovra traza è tanto bella che par una Venera.

Fantina Polo

Venegia, MCCCXXIX[2]

A rough translation is as follows:

As many as there are cities surrounded by walls of stone in Tartary, so is the Eastern Ocean filled with islands, with two long peninsulas between which flow rough waters, up to the great peninsula that the Tartar sailors call Ta Can; the first is named Uan Scian . . . the chain of islands To Qiú, the next peninsula Ta Can that is joined to . . . a bare land; there, the men have swords of ivory and every woman . . . is as beautiful as a Venus.

Fantina Polo

Venice / 1329

The peculiar place-names of Uan Scian, To Qiú, Ta Can, and Fusan may provide a key piece of evidence in deciphering these maps. At first glance, they seem to be nonsense—but in fact they are Italianate renderings of Chinese terms. There is an old Chinese legend of a place called Fusang (扶桑), a land far across the seas. In that legend, these other Chinese places names also appear. This means that whoever made these maps knew of this Chinese story and was able to render the toponyms in Italian. We will examine this intriguing connection between these Rossi maps and this tale in our concluding chapter here.

On the verso of the "Fantina Polo Map 2" are the words "Tabula Geographica Fascio 96. Folio 254." On the verso of the "Fantina Polo Map 1," as noted earlier, we read the inscription "Tabula Geographica

Fascio 96. Folio 255." The "Moreta Polo Map 1" has on the verso the words "Tabula Geographica Fascio 96. Folio 256." These may be clues to the chain of ownership of these strange and intriguing documents and indicate one of the potential directions for future research. From the terms used—for example, *fascio* ("sheaf")—it seems as if these documents were part of a much larger collection at some point.

Grid Maps, Ptolemy, and More Chinese

The name of Moreta Polo, Marco Polo's youngest daughter, appears on several documents in this collection. The first, the "Moreta Polo Map 1" (see pl. 8), displays the same kind of gridded configuration that is found in the "Fantina Polo Map 1" (pl. 6). The second map, the "Moreta Polo Map 2" (see figs. 4a and 4b), has some similar features; that map is now missing, but it is reproduced in Bagrow's 1948 article.[3]

The "Moreta Polo Map 1" is a map covering Europe, North Africa, and Asia; it is signed "Moreta Polo" with the year "1338." Although this map has what appears to be a "longitude-latitude" grid, it is not clear whether this grid was critical in the map's construction. The grid comprises 22 longitudinal sections. In addition to what looks like a Ptolemaic *oikoumene*, we have the addition of the eastern coasts of Asia, and to the far northeast, the label "Transerica Pons" along the coast. Each "column" of the map is labeled with a Roman numeral VII at the bottom, and with what appears to be a Chinese "seven" (七) at the top. If these "units of seven" are degrees, then the entire map covers 7 degrees times 22, or 154 degrees total, which is less than Ptolemy's extension for the *oikoumene*. There are also some crudely written Chinese characters along the right side of the map; these appear to be almost identical to the characters that are found on the "Moreta Polo Map 2."

At the bottom of the "Moreta Polo Map 1," there is a text that begins: "Journey made by Maffeo, Nicholau [*sic*], and Marco Polo from Venice to Acre in Persia, and from Acre finally to Campalu in Tartary, voyaging through valleys and mountains, and finally, to the completely walled city." However, the map does not seem to illustrate the voyage of the Polo clan. The text goes on to state that Marco Polo

drew a Summu Nauticu and navigated the sea up to the eastern extremity, and finally a long peninsula completely surrounded by sea wolves [*luvi marini*] and connected to a huge land, completely unknown that is situated beside the land that by the Moor is called Serica. There [i.e., in that huge land], every woman is [an] archer and each man [is an] archer of the king. . . . There, because of the cold they live in round chambers. The king of [this] locale invited Messer Marco into his chamber. . . . Messer Màrco brought a mask of gold to the king in Tartary, where the king placed him to govern one of his provinces. It is sufficient to note that according to the Syrian sailor in the service of Master Marco Polo, it is said in truth that the aforementioned peninsula is situated DCCCC [= 900?] degrees [or] eight hours difference from the islands off Mauritania in the Atlantic Ocean.

The "Syrian sailor" is probably the "Sirdomap" referred to in the other Rossi documents discussed earlier. It is interesting how the text both does and does *not* relate to the map above it. We do indeed see the peninsula in the seas beyond Asia; however, there is nothing in the text to explain the term "Transerica Pons" that appears on the map, nor is there any explanation of the Roman numeral VIIs that run along the bottom of the map and the accompanying Chinese numbers at the top. The mention of archers is found, too, in the "Bellela Polo Chronicle," examined earlier. That "Chronicle" also mentions that Marco Polo was "entrusted . . . with a mask of gold for the king of the Tartars."

The document has a few final lines of somewhat ambiguous text below the passage just cited: "And more than two hours from Tartary, from where in times past this people emigrated. The writing on the stones says that this people emigrated from Tartary when the gentleman was already there [the meaning here is not clear]. The aforementioned people in the Mongol language call this land Foxaa, [or] more exactly, Focaa." Obviously, this is a reference to the farthest eastern regions depicted in the maps, as well as the legend of Fusang. But, as noted, the map does not actually seem to show the travels of Marco Polo, nor does it bear the place-names that appear in *Il Milione*. Instead, it utilizes Ptolemaic terms. The phrase "when the gentleman was already there" is very unclear, but seems to be the accurate rendering of the Italian here, which is "quando signore zaa sia," with *zaa* being dialectical (apparently, Venetian) for *già*. Perhaps the *signore* referred to here is Marco Polo.

The "Moreta Polo Map 2" has two cartographic renderings: on the recto side is a map of South and East Asia, with an oval cartouche containing an inscription in Italian (fig. 4a); on the verso is a map covering Europe, North Africa, and Asia, all in a "longitude-latitude" grid (fig. 4b), as we observed in the "Moreta Polo Map 1" (pl. 8). This verso map has general Ptolemaic features, but there is also the unusual addition of the eastern coasts of Asia, as well as Cipangu (i.e., Japan)— again, neither of which appears in Ptolemy. We find again in the northwest corner of the map "Transerica Pons," along with a "Sirenus M[are]." Below that region we see a series of islands curving toward a "terra inco[gnita]" that runs along the right hand edge of the map.

As in the "Moreta Polo Map 1," the rendering on the verso side of the "Moreta Polo Map 2" has a grid, where each "column" is labeled with a Roman numeral VII at the bottom, and a Chinese "seven" (七) at the top. There seems to be no explanation for this "seven"; such a unit of geographic measurement does not seem to appear in any other early maps or texts.

On the right hand side of this "Moreta Polo Map 2" is a brief Chinese inscription, looking almost identical to that found on "Moreta Polo Map 1." These Chinese characters in the "Moreta Polo Map 2" appear to have been copied by someone unfamiliar with Chinese writing; their meaning is very unclear. Bagrow offers a partial translation from the sinologist Bernhard Karlgren: "have 1,000,000 men . . . to be King over the states of the four quarters."[4] However, very few of the characters in the inscription can be read clearly; one can make out with reasonable assurance only a few: 王, 有, 百, 人 or 入, 之, and 即.[5]

The Question of Antilla

Another distinctive aspect of this collection of maps and texts is the reference to Antilla found in some of the documents. A discussion of the island of Antilla appears on the verso of the "Moreta Polo Map 2"; this island is also mentioned in the "Map of the New World" in the verso text, as well as briefly in the "Lorenzo Polo Chronicle," both of which will be examined later.

Antilla —variously spelled as Antilia, Antillia, Antilla, and so on— appears as a large, rectangular island on series of fifteenth-century maps, sometimes accompanied by a second isle of similar shape. The earliest depiction of the island is on the Pizzigano chart of 1424, where the name is spelled Antilia. This cartographic entity has generated no small share of academic debate: are these lands the whim of an early cartographer? Are they representations based on reports of navigators venturing out into the western reaches of the Atlantic Ocean? Are they evidence of a pre-Columbian landfall by the Europeans in the Americas?

These questions remain open, although a number of modern writers have tried to claim them as resolved. The historian Armando Cortesão, cited earlier, was one of the first to claim that the island of Antilia on the Pizzigano chart indisputably represented part of the shores of North America.[6] The historians who make such claims have had a tendency to compare explicitly early maps and charts to modern works, attempting to match distances between locales, find similarities in renderings and configurations of coastlines, and so on. This is not a productive approach; instead, one should try to understand the early cartographers' particular worldview, and their motivations and rationales for depicting various lands and seas as they did.

The renowned historian of cartography J. B. Harley very accurately pointed out the following: "The usual perception of the nature of maps is that they are a mirror, a graphic representation, of some aspect of the real world. . . . [W]hen historians assess maps, their interpretive strategies are molded by this idea of what maps are claimed to be. . . . Maps are ranked according to their correspondence with the topographical truth."[7] But this quote also implies, then, that early maps are far from being simple representations of empirically gathered information. Geographic conceptions from antiquity, as well as religion, myth, and speculation all played significant roles in the creation of the depictions of lands and seas that one encounters on old maps. These influences should be kept in mind in looking at early map material, and in doing that, we can see here, too, that the appearance of Antilla on early maps remains an open mystery.

Along the left-hand border of the figure on the verso of the "Moreta Polo Map 2" (see fig. 4b), we can make out the words "Ab Antilla ad TRANSERICA DCCCC Grad / VIII ho [= horae?]." The reference to degrees is unclear, but the "VIII ho" would seem to indicate eight hours. In the text on the recto one can make out the following words: "da Antillia cinque Milla lllc octo . . . Milia [?] da Junonia al . . . P." The reference to Junonia reminds us of Pliny's discussion of Atlantic islands; he has both a Junonia and a Junonia Minor.[8] Despite the reference to Antilla, that locale is not depicted here; however, we do find six dots in the western sea, apparently the six "Fortunate Isles" of Ptolemy.

There are two particularly interesting points about the "Moreta Polo Map 2." The gridded map on the verso seems to be Ptolemaic in nature, although, as noted earlier, Ptolemaic maps were apparently not known in Europe in the period in which this work was supposedly created, that is, the late thirteenth century. The map seems Ptolemaic in the sense that it is drawn with a grid system and it bears a number of Ptolemaic toponyms, such as Serica, Cattigara, and Campalu. It also displays a number of unique traits, such as the fact that where Ptolemy's map ends in the east, this depiction continues, showing an eastern shore-line for Asia, and includes the unique toponyms "Sirenus M[are]" and "Transerica Pons."

The figures on the recto and the verso of the "Moreta Polo Map 2" map display a number of other oddities, mentioned by Bagrow; how-ever, one that he did not explore is the very interesting fact that this map mentions Antilla. As noted above, as far as is known, the island—spelled Antilia—first appears on a map in 1424, the "Pizzigano Chart of 1424."[9] While Cortesão seems correct in his assertion that the Piz-zigano chart is the earliest cartographic depiction of Antilia, it may not necessarily be the first *mention* of the island, as he contends.[10] On the recto of the "Moreta Polo Map 2," the map is signed "Morett[a] Polo"; this daughter of Marco Polo died about 1375. So it would seem that this appearance of the toponym Antilla predates that of the Pizzigano chart by several decades. Bagrow's 1948 article on the Polo maps predates the well-known works of Cortesão on Antilia; Cortesão's earliest writings on the subject appeared in the 1950s. In his study of the Polo maps, Bagrow makes no reference to this island, however, nor does he ex-plore possible origins of the toponym or why it appears on some of the Polo documents. Cortesão, in turn, makes no mention of the Bagrow article.

If we postulate for a moment that this "Moreta Polo Map 2" is some kind of relatively modern fabrication, then we are presented with sev-eral challenges. The fabricator would not have had the Cortesão mate-rial to work with, since we know that this map was presented to the Library of Congress for examination well before the 1950s. William H. Babcock's article entitled "Antillia and the Antilles"—a good, early study of this cartographic entity—was published in 1920, so it is pos-sible that his article could have been used as a source. But there does not seem to be any strong basis for this conclusion; in particular, the way Antilla is used in the Polo maps—in the context of a distance mea-surement—seems quite unique. Any fabricator would have had to have had some earlier source for the utilization of Antilla in the construc-tion of the Polo maps.

Prior to the work of Bagrow and Cortesão, of course, we have the appearance of Antilia in a series of maps, such as the Pizzigano chart. We also have textual references, such as that found in the sixteenth-century work of António Galvão.[11] Only three early sources, however, treat Antilia in the framework of longitude, the manner in which we find it on the "Moreta Polo Map 2." Moreover, only one of these sources puts Antilia in a Ptolemaic context: a map mentioned by the cosmographer Pedro de Medina in his book *Libro de grandezas y cosas memorables de España*, published in Seville in 1548.[12] There we read:

Antilia. Not very far away from this island of Madeira, there is another one which now is not seen. I found this figured on a very old sea chart, and now as there is no news of it, I sought through many ways whether there might be some notice or document, and in a Ptolemy addressed to Pope Urban, I found this island indicated. . . . This island, according to the representation of the chart, is eighty-seven leagues in its largest dimensions, which is north-south, and twenty-eight wide, and it has all over it many harbours and rivers represented. In the Ptolemy that I mentioned, it was situated almost on the Straits of Gibraltar by thirty-six and a half degrees latitude.[13]

In this passage, Antilla is mentioned as being depicted on a Ptolemaic map. In addition, the date of the work would seem to be the fourteenth century—Cortesão suggests that the pope cited is Pope Urban VI, who died in 1389; if such a dating is correct, it puts this "Ptolemy addressed to Pope Urban" very close to the period of the "Moreta Polo" maps.[14] The key point here, too, is that both the "Moreta Polo Maps" and the map discussed by Medina speak of Antilla in a Ptolemaic context. We might note, interestingly, that the Pizzigano chart—where Antilla first appears—is *not* Ptolemaic.

The remaining two sources that treat Antilla in the context of longitude are also somewhat uncertain matches with the material in the "Moreta Polo Map 2." Abraham Zacuto (1450/145?–after 1522) was a famous Jewish astronomer who worked in the court of King João II of Portugal. He may have been discussing the longitude of this peculiar island when he set an unnamed isle at 17½ degrees west of the Canaries.[15] The passage in question comes from Abraham Zacuto's *Hajibbur Hagadol*, otherwise known as the *Almanach Perpetuum*.[16] In chapter 9 of that work, a section that deals with longitude and latitude, we find the following: "It has been understood that the longitude of the cities is calculated from west to east, and, as we are nearer to the west, we have started calculating from the western side. Others calculate the longitude of the cities from an island existing in the west, which stands at 17½ degrees, and from this the majority of the tables run into error."[17] However, Zacuto's description does not seem to refer to a Ptolemaic map, nor does this longitude figure here match anything in the figures for Antilla on the "Moreta Polo Maps."

We should also look at the 1474 letter of Paolo Toscanelli (1397–

1482) for a possible connection to this Polo map. Toscanelli, a Florentine physician and cosmographer, supported the idea that East Asia could be reached by a westward voyage from Europe, more quickly than by rounding the Cape of Good Hope and crossing the Indian Ocean. Toscanelli wrote a letter describing this idea and included a map, which reached King Afonso V of Portugal in 1474. The text of Toscanelli's letter has survived, but not the map—however, the map has been reconstructed, based on the details from that text. The part of the letter that mentions Antilla runs as follows: "And from the island of Antillia, which you call the Island of the Seven Cities, to the very famous island of Cipangu are ten sections, that is 2,500 miles. That island is very rich in gold, pearls and precious stones, and its temples and palaces are covered in gold. But since the route to this place is not yet known, all these things remain hidden and secret; and yet one may go there in great safety."[18] Note that Toscanelli discusses the longitude of this island in terms of its distance from Japan, while the "Moreta Polo Map 2" treats the longitude of Antilla in terms of its distance from Transerica—a region apparently not found in any other source. Furthermore, the Toscanelli letter describes an *oikoumene* of some 230 degrees in extent, and gives a general depiction of lands and oceans quite different from what we find in the "Polo" map.[19]

The mention of the "Island of the Seven Cities" is a reference, mentioned on the Martin Behaim globe of 1492 and other sources, to an old tale of an of an island that "was inhabited by an archbishop from Porto in Portugal, with six other bishops, and other Christians, men and women, who had fled thither from Spain, by ship, together with their cattle, belongings and goods."[20] But even though accounts of this island put the date of the journey in the eighth century, there is no actual trace of the tale before the fifteen century, and thus, it still does not predate the "Moreta Polo Map 2" renderings.

We are left with three possible conclusions. First, there is the possibility that the figures on the "Moreta Polo Map 2" are pure fabrications, with the Antilla distances completely made up, and not based on any source, with just the island name itself taken from some earlier text or map. Next, there is the chance that, again, the map is an early fabrication, with the Antilia distances taken from some contemporary—but now lost—source.

Finally, it could be that the maps are genuine, from the period of Marco Polo, or are accurate copies of ones of that time. In that case, the maps must be based on Antilia material unknown to us, as the description of the island in the "Moreta Polo Map 2" does not match any surviving textual or cartographic depiction. This conclusion would also point to a thirteenth-century knowledge of Antilia; recall that as far as current historical knowledge tells us, the earliest surviving mention of the island is Pizzigano's chart from 1424.

It is interesting to note that the "Marco Polo Maps" here that mention Antilla do not discuss the nature of the island itself. On the Pizzigano chart and other maps that have Antilia, we see depictions of the island itself, or, as in the passage from Pedro de Medina, we have a description. On the "Moreta Polo 2" map, Antilla only serves as a distance marker.

The toponyms on verso figure of the "Moreta Polo Map 2" are India Gangem, SERICA, Campalu, SINAE, Cattigara, Cipangu, Magnus M[are], Sirenus M[are], Transerica Pons, and finally TERRA INCO[GNITA] at the far right end of the map. The map on the recto has a somewhat awkward text in Italian that reads: "The voyage from Venice to Acre in Persia, and from Acre to Campalu undertaken by land and sea by Maffeo and Nicholau Polo and son Marco, Venetian merchants and sailors, and together [with?] that [voyage] of Marco in the Far East . . . that remote empire distant from Antilla five thousand . . . [miles?] . . . taken from the letters of my father. Moretta [sic] Polo. Venice."[21] Note that the opening phrases here are very similar to the text found on "Moreta Polo Map 1." Beginning in November of 1271, Maffeo and Nicolo Polo, along with Nicolo's son, Marco, went from Venice through Acre, coming to the abode of Kublai Khan in 1275.[22] The above passage seems to suggest that Marco made additional travels *in extremo Oriente*; however, the traditional narrative indicates only that he was sent by the Khan to Tibet, the Indian Ocean, and other parts of China. On his way back to Venice, where he arrived in 1295, he stopped in Sumatra, southern India, and Persia. Again, there is no mention in *Il Milione* of travels to distant lands in the northeastern seas beyond China.

For a moment, let us return to the inscription on the verso of the "Moreta Polo Map 2": "Ab Antilla ad TRANSERICA DCCCC Grad / VIII ho [= horae?]." The region of Transerica Pons is found in the northeastern corner of the map. The "DCCCC Grad" would seem to indicate 900 degrees, but that is a rather peculiar figure and, regardless, would be more appropriately written with Roman numerals as CM. The "VIII ho" perhaps stands for "eight hours," or 120 degrees. The initial part of the inscription can be translated simply: "From Antilla to Transerica." We are left, then, with the suggestion that the distance from Antilla to Transerica is 900 degrees (?), or eight hours. The latter figure would indicate, as we have said, 120 degrees.

Once again, the Rossi documents present us with many questions and are subject to a variety of possible interpretations. The puzzling mix of Ptolemaic toponyms, Chinese characters, and text passages concerning various sojourns by Marco Polo in the "Moreta Polo" maps are difficult to interpret. But we can say one thing with certainty: whoever made these maps had some familiarity with the history and supposed location of Antilia—either in the thirteenth century or more recent times.

FIVE

Chronicles and Histories

Two documents in the collection seemed to have been written to elucidate the nature of the collection as a whole, while also providing details on some of the maps and cartographic renderings. The two documents themselves—the "Lorenzo Polo Chronicle" and the "Spinola Chronicle"—do not include any maps or other illustrations themselves.

An Account of the Polos

The "Lorenzo Polo Chronicle" is a text that discusses, briefly, the Polo family and manuscripts left by a "Rugerio Sanseverino" (see pl. 9). The text is signed "Lorenzo Polo, Protonotario, Cajatia, 1556" at the bottom of the last page, while other passages in text are signed "Carlo Sperano," "Fantina Polo," and "Moreta Polo." The text describes the Polo family, some of the maps, and other matters.

The text is on one side only but appears—because of a series of holes in the middle—as if it were bound in a book at some point in its history. The first part of the text seems to be a copy of an earlier text or letter, since at the end of these first three sections we read the words "Carlo Sperano, Altamura / MCCLXXXIV." At the end of the full text, we find "Lorenzo Polo, Protonotario, Cajatia. MDLVI"; it would therefore seem that Lorenzo Polo was the primary author—and editor—of this document. One might then conjecture that he himself was a member of the famed Polo family, but we cannot confirm this. We know of a Lorenzo Polo who held the title of "Regente del Consejo Colateral" of Naples up to around 1558.[1] The text here gives the author the title of "Protonotary," which was a title given to a senior clerk in certain courts of law; it could also refer to a notary appointed by the Holy See.

The document itself begins with a discussion of the family of Marco Polo; a rough translation is as follows:

[1] In the notes outlined in other pages, there was missing a singular item that I will not pass over in silence. The Polo family included one named Andrea, who engaged the Saracens in naval warfare and brought to our kingdom military honors, to whom virtue brought fortune; who had a wife, Clemendia, daughter of a subject of Charles II of France. As King Charles awarded his compatriots with feudal vestitures, with many castles in our realm, [so] Giovanni de Villaclubai was invested with the title of Baron of his castle in Agnone, where Andrea Polo was lucky and happy, and had fortune . . .

[2] On the day of San Giacomo, the 25th of July 1282, I found myself on the island of Pelagusa, piloting a galley that came upon a large Saracen ship. A Venetian galley piloted by Master Andrea Polo and his brother Gio[vanni?] was seen passing by. We set fire to the Saracen ship, and in fact the Saracens fell into the sea, defeated. On the day of San Bartholomew, the 24th of August, at the invitation of Admiral Rugerio Sanseverino, we assembled at Salerno and were celebrated by the barons and all the authorities.

[3] Master Andrea Polo married a beautiful lady named Clemendia, the daughter of the Baron of Agnone, Giovanni de Villaclubai, and his brother Gio[vanni?] sailed for Constantinople . . .

Carlo Sperano / Altamura / 1284

The document at this point shifts and turns to discuss a map apparently belonging to Marco Polo:

[4] In this memoir of mine, the pen of Moreta Polo has participated and that of her sister, Fantina; in the chronicle of their house, they have written as follows:

[5] Universal depiction of Asia from my father Marco; according to the Saracens, there are 8 hours or 115 degrees. In our language one reads that on these islands, each man [shoots?] darts and arrows [at?] strangers [the meaning is not clear at this point] . . . and it seems to him to be a place wild and without any use.

[6] Depiction of India and Tartary, by Marco Polo, and of the many islands that he explored, for which the Great Khan honored him with rule in a province of the realm. No one had ever navigated to the East; a desert of sand three thousand miles from the Tartar realm. But Master Marco, with ten ships, set sail and went by sea as far as a chain of islands and up to a great peninsula. There, there were caves here and there, and they wear pants and tunics of [sea?] lion skin and deer skin.

Fantina Polo, Venice, 1329 . . .

The document then shifts again, where another author picks up:

[7] To finish the most wise narrative of Fantina, I describe completely the part seen in her map, speaking of what is on the verso of the folio.

[8] The voyage from Venice to Acre in Persia, and from Acre to Campalu by land, made by Maffeo Polo and Nicholau, and son Marco, Venetian merchants and sailors, as well as that [voyage undertaken] by Marco to the Far East. Undertaken from an accessible locale [the meaning is not clear here], [so that] one has full knowledge of the vastness of that remote empire. Distant from Antilla 5,000 miles [*milie*] or eight hours. Design drawn [*tracto*] in the house of my father.

Moreta Polo, Venice, 1338

The narrative here goes off in a different direction, and we assume now that Lorenzo Polo is writing, as it is his name that appears at the bottom of the remaining passages. The text at this point begins with a description of certain locales, related to those mentioned in the "Fantina Polo Map 1" (pl. 6) and the "Fantina Polo Map 2" (pl. 7):

[9] After the writings reported above, [there] follows that of Marco Polo, where among other places explored by him [are those] denoted in the Asian language; first, the peninsula of Uan Scian, the chain of islands Ta-Qiú, [and] the contiguous peninsula Ta Can, which is connected to the sea of Foca [seal], as Polo wrote in the margin "Foca marina" [marine seal]. Here the caves . . . [the meaning is not clear here] . . . the king and warriors wore masks of gold and [had] weapons of ivory, and they were pleasant toward the travelers and gave them fur clothing decorated with pearls that seemed [like those of] Venus.

The author then provides a historical aside, with a specific mention of "nautical drawings" penned by Marco Polo himself, and the famed "golden tablet" given to him by Kublai Khan:

[10] But of these examples already cited, I leave aside many that would fill more pages and that I find tedious. I will speak likewise of that which I noted, that Roberto Sanseverino, Count of Cajatia, who married Elisabetta Montefeltro della Rovere, daughter of the Duke of Urbino, did not write attentively concerning that which pertains to Marco Polo, only noting that which he read: the manuscripts of his ancestor Rugerio Sanseverino, which because of their age one no longer understands. And I judge that if the nautical drawings done by Marco Polo had been accessible to Antonello Sanseverino, great admiral of the realm, that as with his relative Roberto, who married Costanza Montefeltro, daughter of the Duke of Urbino, he would have without doubt increased by great praise the nautical school of Amalfi, rather than [?] [an] Arabic institution [*sic*]. But such parchments passed from Roberto to his daughter Maddalena, who married Giulio de Rossi, and so he had the opportunity to register them among the many manuscripts, and annotate them for my curiosity. And as one understands from the Sanseverino Chronicle, written in the very old characters of the various tales of that time, I transcribe only the annotations of Rugierio Sanseverino that refer to the Order of the Military Tribune, originally written on the tablet of gold, that the Great Khan gave to Marco Polo, in recognition of his explorations in the Far East. The interpretation I write as follows:

[11] Among the prefects who served Kublai Khan, Marco Polo of Venice, legate to the province of Anju by the grace of God, other lands were added through the friendship of the Tartars. Many islands in the most distant east not only unknown but never heard of—you were the first to reach them in the past year, and you deserved the honor; for your bravery, traders and voyagers are obliged to adhere to your judgment and to recognize your authority.

[12] Indeed, I will not fail in my annotations here to mention the inscription that follows from that of the Arabs, wise in the nautical art, who in my judgment were employed by Marco Polo in his maritime voyages, and who up to the time of the famous Semiramide, empress of a large part of Asia, had given names to all the islands in the Indian Ocean in the Arab language, which one sees in Syrian characters in the world map which I discuss, and it is this:

[13] India and related islands according to what the Saracens say; Cattigara in Tartary; island of Zipangu [Japan] and related islands; the Peninsula of the Sea Lions; islands adjacent to the Peninsula of the Stags, which is situated four hours difference from the walled provinces of Tartary.

The text concerning the "tablet of gold," above (see paragraph [11]) seems to relate to the "golden tablet" text in Latin that we examined earlier. The text in paragraph [13] here, concerning the depiction on a world map, is similar to what we read in the "Map with Ship" document (pl. 4), where a numbered key has the following:

I. India and adjacent islands, according what the Saracens say. II. Cattigara of Tartary, islands of Japan, and adjacent islands. III. Peninsula of the Sea Lions. IV. Islands connected to the Peninsula of the Stags, situated at four hours difference from the walled provinces of Tartary.

However, the author of the Lorenzo Polo Chronicle comments that the map he is discussing is a "world map" (*mappamondo*), which the "Map with Ship" is not. Moreover, the Chronicle states that the names of the islands in the Indian Ocean on the map are "in the Arab language," which is not the case in the "Map with Ship."

The author of the "Lorenzo Polo Chronicle" ends as follows:[2]

[14] Finally, I conclude [by noting] that I have started the composing of a genealogical tree as evidence of the family produced from this lineage, [with information] accurately taken by me from original writings. These preserved [writing] are of the branch of the Polos [who are] natives [*oriundi*] of Venice.

Lorenzo Polo, Protonotary, Cajatia, 1556.[3]

The internal claims of this document are clear—but how should we approach them?

Tracing Names

This whole text is filled with intriguing details. The narrative tells us that there was a Giovanni de Villaclubai, who received the title of "Baron of Agnone." We do indeed find a family named Villacublai-Sangiorgio, which took control of the town of Voltumo in Caserta after 1290. We also read in the text above that Andrea Polo—Marco Polo's grandfather—married a woman named Clemendia, but other historical sources do not seem to mention this.

In section [10] of the text, too, several members of the Sanseverino family are discussed. We can find corroboration for some of these figures in existing sources. The Roberto Sanseverino mentioned in the text may be the same as the Roberto Sanseverino (1418–1488) who married Elisabetta, daughter of Federico di Montefeltro, Duke of Urbino; he later married Lucrezia Malvotti, who was of a noble family in Siena.

Roberto was First Count of Caiazzo and Prince of Salerno. He was the nephew of Francesco Sforza, the Duke of Milan. Roberto came from one of the most noble families in the Kingdom of Naples. His father was Leonetto Sanseverino.

Roberto had several children; of prime importance here is his daughter Maddalena, who married Giulio Cesare de Rossi. According to the *Enciclopedia Storico-Nobiliare Italiana*, Roberto fought alongside his uncle, Francesco, and after the latter died, he governed Milan. Roberto died in battle at the age of seventy. It is this critical juncture in the family history that may explain the possession by the Rossi family of these documents. Recall that Marcian Rossi claimed a direct connection to Giulio Cesare de Rossi.

We also read of a Rugerio and an Antonello Sanseverino in paragraph [10] of the "Lorenzo Polo Chronicle." This Rugerio may be Ruggiero (II) Sanseverino, also known as Ruggero (II) Sanseverino; he was born about 1237 and died in 1284. He was the Second Count of Marisco. The Sanseverino family was one of the seven great houses of the Kingdom of Naples. The founder of the clan was Turgisio, whose royal blood came by way of the dukes of Normandy. These dukes had come to Naples in 1045 and had obtained the earldom of Sanseverino—hence the family name.[4]

Ruggiero Sanseverino was entrusted to Pope Innocent IV (elected Pope in 1243) as a child. Ruggiero served Charles d'Angiò (i.e., Charles d'Anjou or Charles I) when the latter became the King of Naples and Sicily in 1266. He distinguished himself in the Battle of Benevento, and the King then bestowed on him the title of "Count of Marisco." Ruggiero was married twice: first, to the daughter of the Count of Lavagna, and later to Teodora d'Aquino. They had a son, Tommaso (II), named after Ruggiero's father, Tommaso di Sanseverino (1187– 1246), who apparently had been the first Count of Marisco. Tommaso (II) went on to become the Third Count of Marisco.

Marco Polo returned to Venice in 1295, according to the existing narrative of his journey. By this time, Ruggiero would have dead for eleven years, so we are left with a problem: either this is a different "Ruggiero Sanseverino" from the one mentioned in this text, or some of the dates are incorrect.

Antonello Sanseverino (d. 1499) married Costanza, the daughter of Federico di Montefeltro, the Duke of Urbino. He was Second Prince of Salerno and Twelfth Count of Marisco. Antonello held the title of Grande Almirante del Regno, with Regno probably referring to the Kingdom of Naples. He was the only son of Roberto. Antonello led a conspiracy of barons against Aragonese domination and was forced to flee for a time.

The text of the "Lorenzo Polo Chronicle" is intriguing, but there are many questions. One might think that the detailed information in this text—toponyms such as the peninsula of Uan Scian and Ta Can, the discussion of maps made by Marco Polo himself, and so on—would be found in the traditional Polo narrative. Yet this information does not appear in the narrative, and so the content of our text above strikes the reader as especially peculiar.

Even though the later version of the Polo narrative found in the *Navigationi et Viaggi* of Ramusio has some extra details, that work, too, does not mention the stories or toponyms found in the "Lorenzo Polo Chronicle."[5] The last paragraph of the text here mentions the Polo family of Venice, and the plan of the author (i.e., Lorenzo Polo) to trace the family tree. Existing sources provide us with very little information on the Polo family, and certainly no mention of a Lorenzo Polo.[6]

A Look at Ramusio

We might think that the text here has no connection to what we know—from the traditional sources—about Marco Polo and his travels. But there are some subtle parallels. For example, the version of the Polo narrative in Ramusio has some peculiar extra details not found in the other versions of the Marco Polo narrative.[7] In Ramusio's version, we find a brief discussion of the distant eastern regions of Asia. In 1558, Ramusio published the second volume of his *Navigationi et Viaggi*; there, he dealt with the Far East and included an account of Marco Polo's travels. There is a passage concerning Polo in this Ramusio edition that seems to hint at the Pacific coasts of North America, with a mention in the last line that there seems to be in these distant regions *un altro mondo*. This seems to be the closest connection one might find between Ramusio's text and the depiction given in the Rossi maps and texts:

Departing from the port of *Zaitum*, you sail towards the west, somewhat southwest, 1500 miles, traversing a gulf named *Cheinan*. This gulf is so long that it takes two months to cross it, sailing towards the northeast. Towards the southeast, it washes the entire part of the province of *Mangi*, and on the other side, *Ania*, *Toloman*, and many other provinces I have mentioned previously. Within this gulf there are an infinite number of islands, almost all well inhabited, in which there is found a great quantity of gold of *paiola* [literally "straw"; here the phrase seems to mean "gold dust"], which they collect from the water of the sea where the rivers empty into it. Besides this, copper [or brass] and other things are found, and commerce is carried on with what occurs in one island and not in another. They also trade with those on the mainland, selling gold, copper, and other things, and buying from them what things are necessary to them. In the majority of these islands, much grain grows. This gulf is so large, and so many people live in it, that it seems like another world.[8]

The mention of *oro di paiola* (gold of *paiola*) also appears in the regions of Asia depicted in the fifteenth-century map of the monk Fra Mauro.[9] To the west of his province of Mango, there is the province of Choncha, as far as the sea of Breunto; near this is the inscription: "Qui se trova ora de *paiuola*," that is, "Here one finds the gold of *paiuola*."[10] The situation could be analyzed more readily if the information here in Ramusio concerning the exploration and cartography of northeast Asia were all found in the traditional narrative, Marco Polo's *Il Milione*—but it is not. So, too, then the contents of these Rossi maps and documents strike the reader as especially problematic.

The Ramusio passage above is part of a discussion in the narrative that deals with the seas around Zipangu, that is, Japan. The German historian Christian Sandler noted that the information provided by the text is sufficient to put together the depiction we find in the 1566 Paolo Forlani/Bolognino Zaltieri map (figs. 9a and 9b).[11] A 1561 Gastaldi work has Mangi, and a sea of the same name. There is also the isle of Giapan, and north of that a Golfo Cheinan, and then the province of Ania. To the northeast, in the *altro mondo*—the "other world" of Ramusio—there is the land of Toloman. Sandler also pointed out that it seems that the Venetian cartographer Gastaldi is the first to make this representation, that is, a map with land beyond the farthest northeast reaches of Asia.[12] Gastaldi first introduced this depiction in 1561 and repeated it in a small geographic work of about 1562. In this work, he describes a world map and the configuration of the continents, while putting aside the depictions of Ptolemy.[13] The fact that a work of this Italian cartographer would have such a strait between Asia and America is particularly noteworthy in view of the fact that he himself earlier had put forward a completely different view—in Gastaldi's 1546 "Universale" map, North America links up with Asia along the 40 degree northern parallel.[14] Other Italian cartographers also had such a link.[15]

If one were to accept the evidence of the maps here in the Rossi Collection, they could represent an earlier depiction of this idea of a separation of the landmasses, that is, the northeast reaches of Asia, a

9a   The 1566 map of Paolo Forlani/Bolognino Zaltieri. (Source: Barry Lawrence Ruderman
     Antique Maps, http://www.raremaps.com/gallery/detail/31104/Il_Disegno_del
     _discoperto_della_nova_Franza_il_quale_se_havuto/Forlani-Zaltieri.html.)
9b   The 1566 map of Paolo Forlani/Bolognino Zaltieri: detail of strait between Asia and
     North America.

strait, and then the coasts of North America. The maps in this collection, however, do not employ the terms "Anian," "Toloman," and so on, and so it seems that their depictions are not based on a reading of the traditional Polo narrative and were not the sources for Gastaldi's conceptions.

Sandler argued that the description found in Polo accurately describes the regions of Kamchatka and the Kurile Islands, adding that Polo may have received news of the farthest northwestern regions of North America.[16] This would imply a closer connection between the Polo narrative and the peculiar depictions we find in the maps of the Rossi Collection, with their representations of the farthest northeastern reaches of Asia and part of the New World. While it is impossible to confirm Sandler's conjecture, we can at least see how Marco Polo's narrative could have led Gastaldi—or someone before him—to draw these regions in this way.

In fact, it is somewhat odd that the Rossi maps and texts concerning the furthest reaches of Asia never mention Anian, Toloman, or any of the other geographic information on the Far East found in the traditional Polo narrative. Instead, the toponyms on the Rossi maps tend to include a rather limited collection of place-names from Ptolemy and from Pliny, and several from Chinese lore—with only a few from Marco Polo's account.

In terms of cartography found in the Rossi documents, the "Map with Ship" bears some similarities to the concepts of Gastaldi, with a narrow strait separating Asia and a land beyond. We find other representations that are also somewhat similar to that found in the "Map with Ship"—for example, in the 1566 Forlani/Zaltieri map. An even closer, although rough, match to the depiction in the "Map with Ship" is found in a map by Aaron Arrowsmith (see fig. 10). This map, bearing the long title of "A chart of the Strait between Asia and America with the coast of the Tschutski laid down from astronomical observations made in the Icy Sea during the years 1786 &c. to 1794," was published in 1802. We can compare the regions around the Bering Strait as shown in that map with the same area in the "Map with Ship" (see fig. 11).

It is not clear, however, if these later maps were working from the same sources as the renderer of the "Map with Ship"—or if the "Map with Ship" is a late copy of these maps. None of the well-known depictions of the Bering Strait region—or the earlier, speculative maps of the strait between Asia and America, such as that by Gastaldi—match exactly what we find in the "Marco Polo" materials under study here. Whoever made the "Marco Polo" materials probably was not working

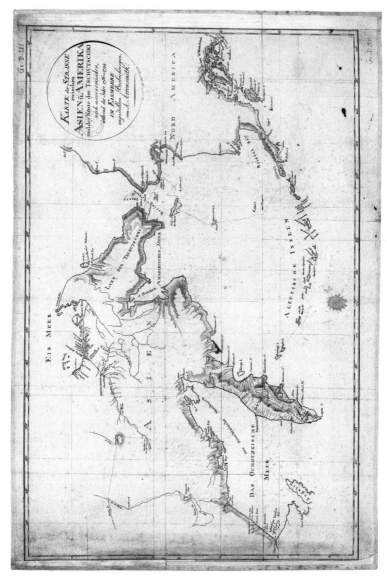

10 A German edition of the late eighteenth-century "Chart of the Strait Between Asia & America . . ." by Aaron Arrowsmith. (Source: Bibliothèque nationale de France.)

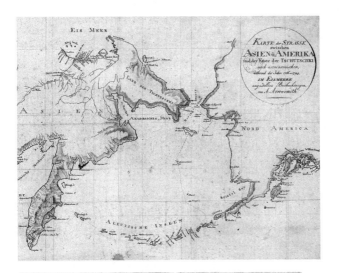

11   A comparison of the regions around the Bering Strait in the Arrowsmith map (*top*) with the same area in the "Map with Ship" (*bottom*).

directly, at least, from these cartographic depictions—the sources were some other maps, now lost, a textual account, or some kind of a report of actual navigations.

We should note, however, that some other works of the sixteenth and seventeenth century directly link Marco Polo with the New World. For example, in a work by the famous geographer and globe-maker Johannes Schöner (1477–1547), an explicit connection is made:

Amerigo Vespucci, who sailed to the west from Spain, surveying the maritime portions of India Superior, believed that part of India Superior to be an island, and instructed that it be called by his name. Now, other more recent hydrographers, coming from other parts beyond that land, have found it to be the continent of Asia, for thus, moreover, they arrived at the islands of the Moluccas in India Superior. Before our time, this continental portion of upper Asia, beyond [the area described by] Ptolemy was explored by Marco Polo of Venice. . . . [And] the land which is called Mexico or Temistitan, in India Superior, which before was called Quinsay, that is, "City of Heaven" in their language . . .[17]

The regions which are beyond the limits found in the description of Ptolemy have not up to now been recorded with certain authority nor described with much diligence. . . . Beyond the Sinae and the Seres, and beyond the 180 degrees of longitude to the east, many regions were reported by one Marco Polo of Venice, and others, but now by Columbus of Genoa and Amerigo Vespucci the shores of these places were surveyed by [voyages] from Spain through the western ocean. . . .

But, in truth, from the most recent voyages made in the year 1519 after Christ, by Magellan, leading ships . . . to the Moluccas Islands, which others call the Maluquas, situated in the farthest east, they found that land to be the continent of India Superior, which is a part of Asia. . . . There are, moreover, these regions of that part: Bachalaos [modern Newfoundland] . . . , the land of Florida, the desert of Lop . . . , the province of Tamacho, Sucur, Sampa or Zampa, Cavul, Tangut, Cuschin, Cathay . . . , and the kingdom of Mexico, in which [there] is the great city of Temistitan, situated in a great lake, but among the ancients was called Quinsay.[18]

This "placing" of Marco Polo in the New World—with the conflating of Temistitan (i.e., Tenochtitlan, the site of modern Mexico City) and the Chinese city of Quinsay—comes, in part, from the connecting of Asia and America in many early maps. This setup is found in Schöner's 1533 globe; the passage above comes from *Opusculum Geographicum*—a work of the same year that, in fact, explains the globe's designs. In this configuration, North America and Central America become an extension of the Asian landmass. Therefore one ends up with a certain—

admittedly odd—cartographic logic, where Marco Polo is described as having explored even these very remote regions. What is yet more peculiar, however, is that the Rossi maps do *not* show this configuration, but rather suggest that Marco Polo reached the extreme eastern edge of Asia, found ocean there, and then engaged in some tentative exploration of islands and mainland beyond.

Frei Gregorio García has the following in his *Origen de los indios del Nuevo Mundo e Indias Occidentales* (1607), in a curious chapter entitled "Marco Paulo Veneto: si estuvo en Mexico . . . :"[19]

And if someone might say that it is not wonderous that the ancients knew of these regions, as they were so far, and so remote, I would respond that as Ptolemy knew of the kingdom of China and makes mention of it in his *Cosmography* . . . when he divides the earth, giving to Asia, and to the other parts, the provinces and lands which belong there, and stand below his mainland; it matches with New Spain, being Tierra-firme up to the canal or strait that is found between the kingdom of Anian and Gran Tartaria, and China stands well near the aforementioned kingdom, as one can see in the Terrestial Globe and the General Map of the World of the most modern [cartographic works], particularly in that of Pedro Plancio, or that of Enrico Alangren. {And more clearly in the sphere which Fr. Francisco sent to the Archbishop of Palermo with a letter, in which he deems the city of Themistetàn to be the same as Tenuchtitlàn, or Mexico, where, he affirms, there entered many years before Don Hernando Cortes, the famous travelers to the Orient Marco Polo Veneto, John Mandeville, and Fr. Oderico of Forojulio, supporting this idea with various particulars, drawn from the voyages of these men, and compared with the second *Relacion* of Don Hernando Cortes, believing that one could not conclude it to be any other city but that of Mexico and its empire, riches, and magnificence, that relate to the Great Khan of the Tartars, and [believing him] to be the same as the Khan Montezuma and his court of Themistetàn, reinforcing [this idea] by saying that Asia, by way of Culuacàn, is [part of] the lands of New Spain, a rare and singular thought, with the misfortune of being noticed by few, and followed by none.}[20]

We find the city of Temistetan in the Oronce Finé map of 1531, in the area corresponding to Mexico. It is placed there alongside some of the Asian toponyms of Marco Polo, notably Cathay and Mangi.[21] These texts, like the work of Gastaldi, exhibit an interpretation of the Marco Polo's Asian narrative so that it comes to include certain regions of America, but in a manner rather different from that of the Rossi maps.

Another Chronicle

The "Spinola Chronicle" contains a text on the recto in the form of a letter to "Elisabetta Feltro della Rovere Sanseverino" and signed "Guido Spinola" (pl. 10). At the end of the text, we also see the words "Cagliari, 25 November 1524." The text here is closely connected with the "Map of the New World," which we will discuss in the next chapter. The "Spinola Chronicle" describes some peculiar voyages and other matters:

Most Illustrious Lady Elisabetta Feltro della Rovere Sanseverino, Duchess of Urbino[:]

Thank [you] for the consideration of Your Excellency concerning the geographic chart of the New World, which we have in part followed as much as was possible in coasting the Atlantic shore. But we were guided by a princess who after some years took a liking to and joined our pilot Gianus Sanguinettus, accompanying us to the royal home of Mahigó [= Mexico?]. There, we were astonished in our admiration by the architectural and sculptural magnificence, but above all by a treasure of the metallurgical arts, consisting of an amphora of gold and silver. There were also statues of astrological idols decorated with precious gems.

The opening of the letter is rather puzzling; who was this princess? Who was this "Gianus Sanguinettus" and what was this voyage? The general idea here seems to be that this Duchess of Urbino gave a map of the New World to the writer of the letter. But the precise nature of this voyage is uncertain.

The letter continues:

The wondrous description of Sanguinettus was sufficient to inspire Captain Ginu Andrea d'Oria, to show [us] a great book, done in vellum, in which was drawn all of the New World; thus we came to understand that it is separate from the great Asiatic land, and revealed to be a true [i.e., truly separate] continent.

This is an interesting idea indeed; historically, the period immediately after the voyages of Columbus was filled with uncertainty as to the exact relationship of Asia and America; were they one in the same place? Or were the Americas a separate landmass?[22] Here in this text, the implication is that this question was, in fact, already answered on a map of the period. However, historically, it seems that at least after his first voyage Columbus still believed that the lands he had reached were Asiatic.[23]

Asia or America?

However, Columbus himself at various times espoused different goals prior to his initial voyage of 1492. Columbus, indeed,

during the period when he formulated his Atlantic designs and sought patronage for it, in the 1480s and early 1490s . . . could have had three possible destinations in mind: Asia, the Antipodes, and as yet undiscovered islands. Historians and biographers have generally been anxious to pin him down to one or other of these. . . . The objective evidence, however, suggests that he considered all three destinations at different times, or sometimes simultaneously, and advocated them severally in addressing different audiences.[24]

The historian Felipe Fernández-Armesto has shown how all three of these different goals were articulated by Columbus in various circumstances. This also demonstrates the variety of worldviews available to those who made up Columbus's audiences, before and after his voyages:

When Columbus returned from his first voyage, despite energetic avowals that he had been to Asia, most Italian commentators seem to have assumed that his discoveries were Antipodean in character: the ready acceptance of this theory in so many sources shows that it must have been current before Columbus' departure.[25]

The "extended Asiatic" world-view, according to which the shores of East Asia were believed to be relatively close to Europe across the Atlantic Ocean, was available prior to the voyage of Columbus. Indeed, Columbus knew about such conceptions from the world maps of Henricus Martellus and the cartographic ideas of Paolo Toscanelli.[26]

This uncertainty as to the nature of the newly found lands—part of Asia or a new continent?—and, again, the resulting uncertainty as to the extent of the world's landmasses, clearly appears in the 1507 world map of Martin Waldseemüller (see fig. 12). This shows a separate landmass, bounded on the east by the Atlantic, and on the west by ocean. However, there is still ambiguity here, as that slim landmass is rendered with a western coastline that is purely conjectural.

12   The 1507 world map of Martin Waldseemüller. (Source: Library of Congress, Geography and Map Division.)

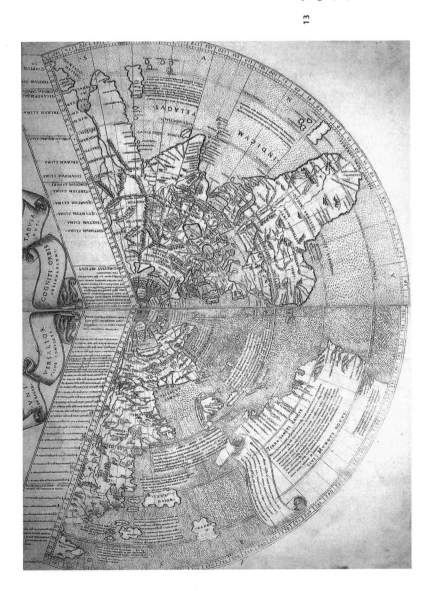

13 The Johann Ruysch map of 1507.
(Source: Martayan Lan, http://www
.martayanlan.com/cgi-bin/image
.cgi?21975.zm.1.jpg.)

In the time of the Colombian voyages, then, there was uncertainty. Some held very particular views, however; for example, certain thinkers of the period believed that the lands recently found across the Atlantic Ocean were definitely Asiatic. Such an idea was put forward by the scholar Peter Martyr and is reflected in a series of early sixteenth-century cartographic depictions such as that in the Johann Ruysch map of 1507 (see fig. 13).[27]

Returning to our "Spinola Chronicle," we read:

An equally wise interpretation was that recounted in the historical book, which affirmed that in ancient times, Asyala, goddess of the marine nymphs, with a retinue of chaste women, in ships controlled by the magnetic needle, passed from the Flavidu Sea into the placid ocean [*oceanu placidu*], having [made] a safe exit to the opposite continent. There, she was joined by her many beloved Anui, and others of the Auzaci mountains, who founded [a] city and castle, and crowned a king and queen.

Again, we are presented with a rather odd story that does not seem to appear elsewhere in the historical record. Several terms are unclear, while others refer to what strike the readers as strange and distant locales. The term "Auzaci," which we find elsewhere in the Rossi documents, comes from Ptolemy. The term "Anui" is uncertain—perhaps a reference to the Ainu of northern Japan, or perhaps to the Anui River in Siberia, but these are pure conjectures. Also, how we might identify the Flavidu Sea is unclear; in Latin, *flavidus* refers to a kind of yellow color. In contemporary cartography, the term "Yellow Sea" refers to the body of water between the Korean peninsula and Mainland China. As for Asyala, her identity is also a bit mysterious; in the text here, she is described as a "goddess of the marine nymphs," a role that in the Greco-Latin pantheon was filled by Artemis/Diana.

The "Spinola Chronicle" concludes:

According to what was affirmed by Genoese sailors who had navigated in the Pacific Ocean [*oceanu pacificu*], [which] lacked dangerous marine seaweed, we were convinced that the ancient Sereci [i.e., the people of Seres, the Chinese], sailed the said ocean unimpaired, and settled on the opposite continent, [a land] abundant with every benefit of creation; all that was impossible in the Atlantic until the bold Genoese navigator Columbus. In fact, not even the zealous pilot Cajus Pisanelus dared to sail beyond the westernmost island of the Canaries, because of the dangerous marine seaweed and the tempestuous vortices of the sea. This took place during the reign of Alfonso V of Naples [1396–1458]. That which raised our anxiety the most was the fact that in the historical annals there was not registered any departure of men to the continent before the Latins, neither in the central, nor in the dark, nor in the southern [parts of the] vast Atlantic.

I trust that these narrated events will please Your Excellency.

Your devoted servant,

Guido Spinola

from Cagliari, on 25 November 1524

This text presents a similar collection of mysterious names and places, some which we see elsewhere in this collection of documents. The Gianus Sanguinettus mentioned earlier is not readily identifiable, nor is Cajus Pisanelus.

As with other documents that we have already examined, however, we can see an interesting pattern: there is a kind of "in and out" weaving of connections between the material in these documents and existing sources such as the Polo narrative, Ramusio's work, the geographic description of Ptolemy, and various late medieval and early Renaissance maps. Many times, perhaps even most of the time, these Rossi documents include cartographic configurations, descriptions, and persons whom we do not find in any existing historical sources.

At other times, we find matches or at least echoes of actual early maps, texts, and so on. In the passage above, for example, there is a clear reference to Alfonso V, a historical figure. Even the reference to "dangerous marine seaweed" is not unusual; we find this in much more mundane early works concerning geography. For example, Avienus, in his *Ora Maritima* from the fourth century AD, talks about the encounter by Himilco, a Carthaginian navigator, with obstructive seaweed in the Atlantic.[28]

In the "Spinola Chronicle," we also see the term "Sereci" used, a term found in classical sources, with "Serica" referring to Asian lands. But the mention of the Chinese having "settled on the opposite continent"—a rather bold statement—makes us wonder, indeed, who has penned this text. Certainly, this statement resonates with Chinese legends, such as that of Fusang, but the exact connection between these all strands of history and myth is unclear.

SIX

Maps of the New World

The "Map of the New World" presents us with further questions but also reveals connections to other documents in the collection. On the recto, we find a map that includes Europe and North Africa, as well as North and South America (pl. 11a). One oddity, however, is that here the North American and South American continents are labeled "Columbia Septentrionalis" and "Columbia Meriodionalis" respectively.

There is a short text below the map, a text that takes the form of a letter addressed to a "Elisabetta Feltro della Rovere Sanseverino" and signed "Guido Spinola"; at the end of the text we read "Cagliari, 20 October 1524." On the verso of this "Map of the New World," we find a text mentioning Antilla and the famed explorer Hernando Cortez (pl. 11b).

Making Sense of the Figure

Immediately, we see connections to the "Spinola Chronicle" (pl. 10). That document, too, is addressed to the same person, and also signed by this Guido Spinola; it also bears the date of 1524. With the "Map of the New Word," we seem to be dealing with a different period entirely, as compared to many of the other maps in the collection. Whereas other Rossi documents, such as the "Fantina Polo" maps (see pl. 6 and pl. 7) and the "Moreta Polo" maps (see pl. 8 and figs. 4a and 4b), seem to have their roots in the thirteenth and fourteenth centuries, the figure on the

recto of the "Map of the New World" is accompanied by an explanatory text in the form of a letter with a sixteenth-century date.

The map here is also quite different in content and form from the
other maps in the collection and presents its own complexities, as we
will see. The map shows Europe, and the northern portion of Africa, as
well as parts of North and South America. The Mediterranean Sea is labeled with the rather archaic toponym Mare Internum. The oceans are
given as Oceanus Atlanticus and Mare Pacificus. Some Atlantic islands
are shown off the northwestern corner of the Iberian Peninsula and off
the coast of Africa.

Of special note here is the large island labeled "Antilla." We may
recall from earlier discussions here that the "Moreta Polo Map 2" also
includes a reference to this island, as does the "Lorenzo Polo Chronicle." In the rendering on the recto of the "Map of the New World," we
find a large island with this toponym. Perhaps by coincidence, or cartographic confusion, an island with a very similar shape—but labeled
"Cuba"—appears in the "Columbus Map," discussed below (see fig. 14).

In the existing early representations of Antilla, such as that in the
Pizzigano chart of 1424, the island is depicted as clearly rectangular

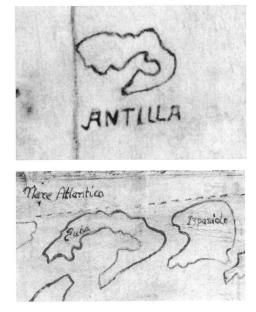

14   A comparison of the island of Antilla in the "Map of the New World" (*above*) with Cuba
      in the "Columbus Map" (*below*).

15   A comparison of the isle of Yezo from the 1630 map of João Teixeira Albernaz (*at left*)
with the island of Antilla in the "Map of the New World" (*at right*).

in form, with distinct bays. By contrast, here in the "Map of the New World," Antilla appears as a small, hooked island in the middle of the ocean. To add further to the puzzling nature of this cartographic artifact, the "hooked" form of the island as shown in the "Map of the New World" seems to appear in a Portuguese map of 1630 (see fig. 15).[1] In that map, we see the island of Yezo, commonly thought to exist north of Japan in this period.[2] In this Portuguese rendering, Yezo looks remarkably similar to Antilla as it appears on the recto of the "Map of the New World," except in reverse.[3]

The western portion of the "Map of the New World" is odd in terms of its rendering of the landmasses of the New World and their names (see pl. 11a). As mentioned earlier, North America and South America are labeled "Columbia Septentrionalis" and "Columbia Meridionalis," that is, "North Columbia" and "South Columbia." These are obvious references to Christopher Columbus, but nonetheless these toponyms do not seem to be found in any other historical sources.

In the middle of the map, we see three vertical lines. These lines are reminiscent of the line of demarcation of the Treaty of Tordesillas of 1494, as shown in the famed Cantino planisphere, but there seems no further connection here.[4] The lines in our "Map of the New World" here may have been placed there to indicate distance: between the lines, at the top and the bottom of the map, we see <I>, which would seem to be the Roman numeral CIↃ, equivalent to 1,000. But this number does not seem to have anything to do with the figures of the Treaty of Tordesillas, which determined the division of Spanish and Portuguese discoveries by a line measured from the Cape Verde Islands.

Below the map, we have a text addressed to a member of the San-severino family:

Most Illustrious Lady Elisabetta Feltro della Rovere Sanseverino,
We have received your letter and understand that which you write, [namely] your desire for a geographic chart [*tabula geographica*] of the New World, and we promptly decided to send one of our men from Corsica to Ancona, [and he] will himself consign to you such a chart. I, and Gianu Sanguinetu, and the captain of the Genoese galleys, Andrea Doria, in the years of trade in African slaves to New Spain, with the help of an Indian princess, had the fortune to sail to the center, then to the south, and then to a good part of the north [of these lands]. On your behalf, rest assured that we have sought to delineate such lands as accurately as possible. To you I offer a hand [i.e., in greeting] from Cagliari, 20 October 1524.
Guido Spinola

This letter is addressed to "Signora Elisabetta Feltro della Rovere San-severino," who is referred to as "Duchessa de Urbino" in another docu-ment in the collection, the "Spinola Chronicle." It is signed by a man named Guido Spinola and bears the date 1524.

Who is this Elisabetta? She may be the same person as Elisabetta della Rovere, first wife of Alberico (or Alberigo) Cibo (or Cybo) Malaspina, the man who helped Giulio Cesare de Rossi obtain a wife. We may re-call from the introduction here that this Giulio Cesare de Rossi was apparently the ancestor of Marcian Rossi, the Italian immigrant who brought these maps to the United States. Elisabetta was the daughter of Francesco Maria della Rovere (1490–1538), the Duke of Urbino. But that Elisabetta lived from 1529 to 1561, and so cannot be the same person as the one addressed in the letter above, since the letter above bears the date of 1524.

As for Guido Spinola, his identity is uncertain. Spinola is a famous name among the Italian aristocracy, with direct descendents still exist-ing today. One of the early members of the Spinola family was indeed named Guido, but he lived some three centuries before this document was written. Once again, we are presented with a number of names—in a complex network of connections. How are these figures to be identi-fied? And does the presence of the names here help us determine the veracity of these documents?

Ventures into the Atlantic Ocean

On the verso of the "Map of the New World," we find another text, rather peculiar in terms of content and structure (see pl. 11b):

And what to say of an island even in ancient times, although opposite [i.e., off the coast of] Mauretania, was occupied by Roman-Etruscan legions? If Your Ladyship [la Signoria vostra] might have sailed in the Atlantic Ocean, she would have admired with compassion the sailors. As Taddeo Visco reports concerning the last island, [it is] called Antilla. There, Cajus Pisanelus recovered in such safety that he called it a paradise in the midst of the dangerous vortices of the sea.

I add how [this] stirred up our desire to visit the capital of that realm. During our stopover on the island of Antilla, [which is] situated between the New Land and Spain, a place where volcanic flames erupt continually and [there are] earthquakes, and which is noted for incursions by pirates. There, Spanish sailors were saved, among them an Indian, who, grateful for our help, showed us the king and the capital of that land. This stimulated our eagerness to explore it.

Perhaps it will be pleasing for you to note, as I continue to write concerning our reaching the capital, which is situated on the southern side of the great gulf of that land, that there the king received us with kindness, and gave to Captain Andrea Doria an amphora of gold. In addition to this, he honored us with a magistrate who rendered our image [i.e., did a portrait] and described our entrance on vellum. We have sent to [satisfy] your curiosity the gold pen with which he described our presence. Later, Doria's amphora of gold of inspired Don Fernando Cortez to visit the capital, and [there] he abused the hospitality with slaughter and plunder.

As in the other documents, this text presents us with a number of puzzles and tantalizing hints. The character "Taddeo Visco" is particularly odd—there seems no other mention of this person in extant historical sources. The figure of "Cajus Pisanelus" also does not seem to appear anywhere else except in the "Spinola Chronicle"; the name looks distinctly classical, and one wonders if the writer of the text here is referring to some otherwise unknown—or imaginary—early Latin source. As for the mention of Antilla, the information in this text does not match what we find in other descriptions of that obscure island. One might speculate that the description of the isle as "situated between the New Land and Spain," and subject to volcanic activity, might lead to an interpretation of the locale as the Azores. But we are left with no clear indications.

The name Andrea Doria, however, is well known: he was a Genoese admiral and statesman. He was born in Oneglia, Italy, in 1468, and died in Genoa in 1560. No existing record concerning Andrea Doria says anything about an "amphora of gold," however.

Finally, we have mention of the famous conquistador Cortez; the text notes his coming to the "capital" and his misdeeds. Historical accounts tell us, in fact, that Cortez destroyed the Aztec capital city of Tenochtitlan in 1521.

An Odd Resonance

Interestingly, the peculiar mention of "Roman-Etruscan legions" occupying an island off the coast of Mauretania—and even the description of it as "a paradise"—*does* indeed find some resonance in another ancient source. Diodorus Siculus was a Greek writer about whom little is known; he flourished some time before 30 BC and wrote what is known as the *Historical Library*, an extensive work on the history of Greece, Egypt, and other parts of the ancient world. In a discussion of islands in the ocean, he writes as follows:

But now that we have discussed what relates to the islands which lie within the Pillars of Heracles [i.e., in the Mediterranean], we shall give an account of those which are in the ocean. For there lies out in the deep off Libya an island of considerable size, and situated as it is in the ocean it is distant from Libya a voyage of a number of days to the west. Its land is fruitful, much of it being mountainous and not a little being a level plain of surpassing beauty. Through it flow navigable rivers which are used for irrigation, and the island contains many parks planted with trees of every variety and gardens in great multitudes which are traversed by streams of sweet water; on it also are private villas of costly construction, and throughout the gardens banqueting houses have been constructed in a setting of flowers, and in them the inhabitants pass their time during the summer season, since the land supplies in abundance everything which contributes to enjoyment and luxury. . . .

In ancient times this island remained undiscovered because of its distance from the entire inhabited world, but it was discovered at a later period for the following reason. The Phoenicians, who from ancient times on made voyages continually for purposes of trade, planted many colonies throughout Libya and not a few as well in the western parts of Europe. And since their ventures turned out according to their expectations, they amassed great wealth and essayed to voyage beyond the Pillars of Heracles into the sea which men call the ocean. . . . The Phoenicians, then, while exploring the coast outside the Pillars for the reasons we have stated and while sailing along the shore of Libya, were driven by strong winds a great distance out into the ocean. And after being storm-tossed for many days they were carried ashore on the island we mentioned above, and when they had observed its felicity and nature they caused it to be known to all men. Consequently the Tyrrhenians, at the time when they were masters of the sea, purposed to dispatch a colony to it; but the Carthaginians prevented their doing so, partly out of concern lest many inhabitants of Carthage should remove there because of the excellence of the island, and partly in order to have ready in it a place in which to seek refuge against an incalculable turn of fortune, in case some total disaster should overtake Carthage. For it was their thought that, since they were masters of the sea, they would thus be able to move, households and all, to an island which was unknown to their conquerors.[5]

A similar story is found in Plutarch, who recounts a story that speaks of Roman soldiers returning from

Atlantic islands . . . two in number, separated by a very narrow strait; they are ten thousand furlongs distant from Africa, and are called the Islands of the Blest. They enjoy moderate rains at long intervals, and winds which for the most part are soft and precipitate dews, so that the islands not only have a rich soil which is excellent for plowing and planting, but also produce a natural fruit that is plentiful and wholesome enough to feed, without toil or trouble, a leisured folk.[6]

The same story is found in pseudo-Aristotle's *On Marvellous Things Heard*; that version is as follows:

In the sea outside the Pillars of Hercules, they say that a desert island was found by the Carthaginians, having woods of all kinds and navigable rivers, remarkable for all other kinds of fruits, and few days' voyage away; as the Carthaginians frequented it often owing to its prosperity, and some even lived there, the chief of the Carthaginians announced that they would punish with death any who proposed to sail there, and they massacred all the inhabitants, that they might not tell the story, and that a crowd might not resort to the island, and get possession of it, and take away the prosperity of the Carthaginians.[7]

Moreover, these ancient stories were in circulation during the period of early Atlantic exploration by the Spanish and Portuguese; they are discussed, for example, by Ferdinand Columbus.[8] It is perhaps not surprising, then, that they appear in this "Map of the New World." But the particular details given in that document are unique.

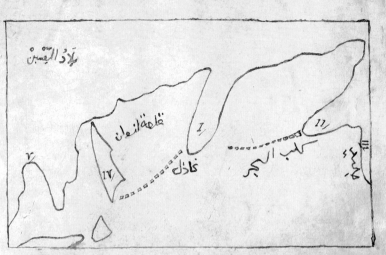

Pzió ke le nobele lezedrixie poran trovar plú dileto zerca el regname del pexu femeninu
en la Eina e lutan levante/ me pare Marco polo conſeglió a me Belleza de moſtrar queſto
mappamondo lo qual otene dal pilota Biaxio Sirdomap lo qual aveua da ben trenta ani
coſtegiato l'Aſia da la Siria enfin all'eſtremo levante marcadando pelixa de phoca marina //
En queſto mappamondo ſe oſcerua cun plú claritade como miſier Marco polo feze vela dal
colfo Mangi e verſo levante enſin a la peniſola deli zervi/ alí encontró el pilota
Sirdomap lo qual depoi lo duxeua à l'iſola dele femene ſita a tramontana e occidente/
E ancora ve digo ke ſegundo como periſe Sirdomap/ en queſta contrá é una zenerazion
ke ſegundo como le dixean li Oracoli,/ en un tempo remoto/ queſta zente á caxon
de la penuria de cibaria/ abandonaron le caverne de li monti Auzei e dala ſcythia
ſcore tuta l'Aſia e entró en queſta contrá ke mó demora // Ogna provinzia é apellata
en loro miſto lenguazo ſcythicu e Tartaro/ perzió Sirdomap nomeava le contrá anco
en lengua Siriaca como quí ſe vede ſcrita // Mai ebe accaſo a la longa eſtreta iſola à la
qual negú omo ní aſiatico ní latino ardí poner paſſo en queſta iſola ſi no iera faitato e mori to da
le arcadrixie en deſeſa de la loro caſtelak e ſu praħumana belexa // Ma miſier Marco polo piando
ke duxeua ſiego una damexela/ poré preſentarſe al palazio de quela reina/ cuppi la ſeva corteſia
à donde piexa prezioſe//

I/ peniſola de li zervi //
II/ peniſola phoca marina/
III/ Vale conzonta e giazata/
IV/ Iſola dele femene //
V/ Eolſo Mangi //

مار ي ثلاثينه سنه سافنسه ع ستا
ال بلاد جليد شفيلي سي جلاو كلد الجب
صالح هاك ك ابنه طق٩ ابوص ييكا لقه محى سلى
شيه ١ طرطط

يا ها سود١ جا٩
أين

١٢٧٧

§ El supra manuscripto datatum CIƆCCLXVII, el capitanio Sirdumap, ha voluto dare poche parole certificando al celebre exploratore Marco Polo del suo trentesimo anno de navigatione per mari dalla Syria el lungo la costa maritima de tutta l'Asia infino all'estremo levante ed una peninsula da ipso appellata phoca marina mercadando pelicie de foca// Come se a bocca ne portasse, ipso scrive con particularitate circa quella remota penisula unde gioia de mari, ove el populo dal freddo extremo che existe, vive nelle caverne// Quel che da piu satisfatione e quanto ha scripto circa lo misto lenguagio Scythico e Tartaro parlato da quel populo che li fa tanto consanguinei dell'una e l'altra stirpe quanto creder se possa// Tutto questo concorde con quanto ipso nappo a Marco Polo, che secundo la credentia delli astrologi locali, in tempi antiqui, li exotici montanari migrarono dalla Scythia e conjunsi a errangi Tartari se stabilirono in quel la regione ove abunda omnia specie de pescatione a alimento, vestimento e metatura// Inferne e degno de laude qui fu ritrovatore del populo Austcu i quali trasperirono a noi l'uso della preciosa pelicia e guidò el nostro Venetiano al piu remoto angolo del globo Terrestre//

niore peralpeppe, ducheppe, marche, conteppe e tute le semene a cui plaxepe de paver
del reame dele femene innela Cina e innel lutan levante poran lezere con propri ocli
quanto á cuitado meo pare e io Bellela ò poxito pagando me á dictato.
Dapó ke mixier Marco polo xera cognoxciuto in tuta Cina per pó fervore, la moier de Fafur,
reina delexemene innela provinzia de Magi le comprdó meppagio per Fupint, reina dele
femene in lutan levante e po pomé al comando de vinti marinari cini e paraxeni e con
granda nava el pexe vela dal colpo Magi e xere la catena depixene xpole ke enxtoapono el
promontorio del lato oxrental de quel colpo e navegó verpo levante, pop intró in mar
oceanu dove preipto levoppe una pí endorbolada unda, ke el crexinu naukeli mutava
de zá de lá forzando navegar al lato tramontana de una catena de xpole la qual
attornearno lo maxo e pe eptende a levante enfin a una pompola ove mixier Marco
polo depepe de nave in tera in ventioto zornate dapó ke l partí da Cina. In quepta contrà
fo paludato da un pixiano de nonne Sirdomap marcadante de pelixra el qual inopixteli
gran alegroza de poa veqnuda: lá el aveva fato marcadanzie de pelixra per ben trenta
ani e plu volte enfin a un'oxra pompola verpo tramontana e levante nomeala phaxa,xi;
la qual ó a doalanto deptantra da Cina. Lá pono zente ke amo lenquazo tartarexe ke apena
algun poxia creder p'el no vedepe: vano veptide de pelixra de phoca marina, vueno de pexcaxon
e pano xape poto tera; pta pompola e conzonta a valle pangone ed é pí gran gxaze, ke per zexlo
uno troveria abippo pe v'andappe. El operpe depxxpo dele pó navigaxion a mixier Marco
polo. Cuppé pledero inpembre per zinque dí in gran alegxeza. Ma un dí mentre tute
reran a monpa, advene ke le duxonto movere del calipo pe approppirnonno epe mepeno
a cantar e ballar. Mó prontoxoppe minaza dele axcadori e tre cini e uno pixiano pono
pantati. Cuppé mixier Marco polo determinó de appexrimentar le piu quatro archlubuxi ke aveva
fato in Cina: lo remor dela xpcopprata ferapuxir li axcadori, dala paggra. Ma lo calipo
pulupepato de tal orxibil arima, xazé a pé de mixier Marco polo e fexel trabuto con zore.
poppa le appedó una mapxara d'oxo per lo xe dela tartari. Ma per caxon de quela
minaza, mixier Marco polo determinó de partir da quela contxà per andar al
zegname de fupint, e domando congeglo a Sirdomap el qual delibexó de andax amEl a quel
zegname. Cuppí con ambe le nave fexan vela verpo tramontana e ponente e in men de
doxe zorxate zonzero a una longa e pxxela xpola e aveno axceppo a uno colpo ove no
troveno zelá, ma troveno un palaggo tuto coverto d'oxo pxippo; in la villa ptava
fupint, penotata pup catxexa d'oxo ben ordenada al comando de una pchiexa de doemila
damixele apelado Bikerne, con lanza in man veptide de pelixra d'ermerlino
ornate de perle e zore: tant reran bele quele damixele ke parova Nymphe.
Quando la reina papé, per ke mixier Marco polo poppe vegnudo, lo xixevó con gran cortepia
e le appedó una lanza d'oxo adoxna de zore per la reina dele femene in Cina. Poppa partippe
da quela xpola e ando per marco verpo ponente. Innela zitá de Quipai trovó mille damixele
con loro reina; equando quepta gran Dama vede mixier Marco polo, avene grande pta
xxere e lo remunexó con molto zore. Ma lo xe e li pui Baroni tartarexi le fen gran
alegrexa e peptintade. (Dhleto pra per via lezetarne. Ita pat)

4   Document 4 ("Map with Ship"), recto. (Source: Library of Congress, Geography and Map Division.)

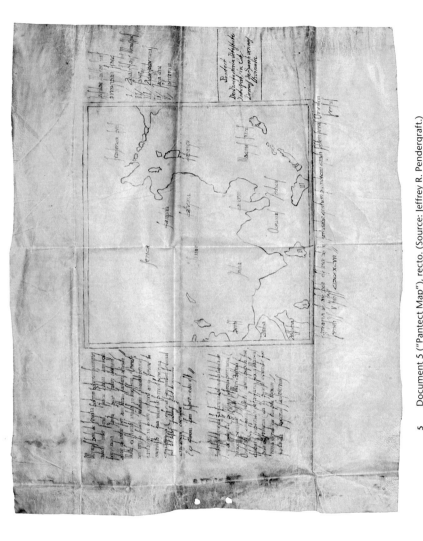

5   Document 5 ("Pantect Map"), recto. (Source: Jeffrey R. Pendergraft.)

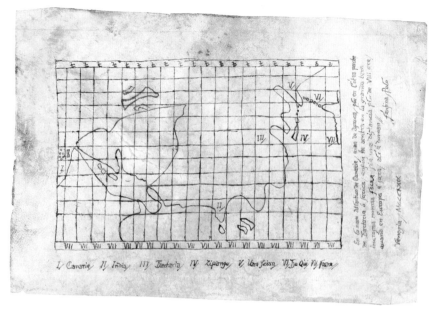

6 Document 6 ("Fantina Polo Map 1"), recto. (Source: Jeffrey R. Pendergraft.)

(de quanta cità é en Barbaria de farti buta murca d'enterno /
tant é pleno lo maro Oriente de Isole cun due longa peninsula
inter li qual li cammino acque entorbaloe en fin en la grandia
penisola ke li nauti Barbari nomean Ba Can / la prima de
morne Uan scian / possa la catena de Isole Do Qiu / la prima
arma Penisola Ba Can ke é controuta a fauza / Dera
derivada // alí é l'Adorno cun spago de avori et rascaduna
permená perra traza é tantobella ke per una Venera //

Fantina Polo

Venegia / MCCCXXXX /

Veti el (osa seguento al numero /
I / Barbaria ó Serica / II / Zipango / III / Uan scian /
IV / Do Qiu / V / Ba Can / VI / facua //
VII / Maro Oriente //

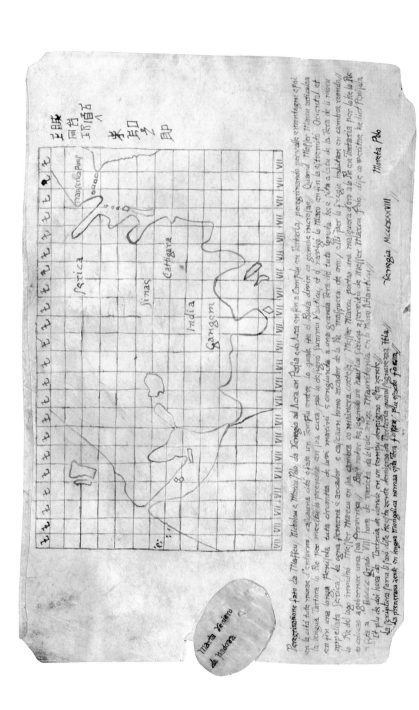

9   Document 9 ("Lorenzo Polo Chronicle"), recto. (Source: Jeffrey R. Pendergraft.)

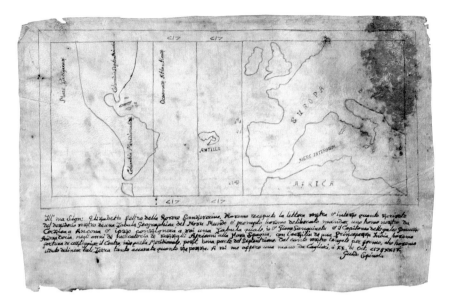

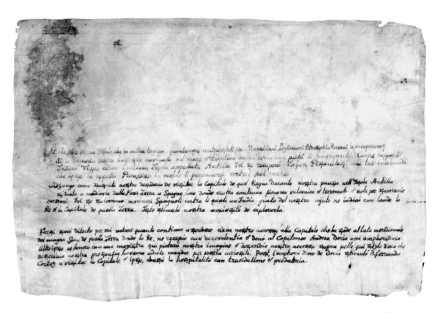

11a Document 10 ("Map of the New World"), recto. (Source: Jeffrey R. Pendergraft.)

11b Document 10 ("Map of the New World"), verso. (Source: Jeffrey R. Pendergraft.)

12   Document 11 ("Columbus Map"), recto. (Source: Jeffrey R. Pendergraft.)

A Reference to Columbus

The "Columbus Map" is another map showing the Americas, but quite different from the piece described above, the "Map of the New World." Here in the "Columbus Map," we have a large rendering of part of the Americas, but delineated in a rather awkward and distorted manner (pl. 12).[9] There is a brief legend in the lower right corner of the recto side, and nothing on the verso.

The legend reads:

Geographic chart [*tabula geographica*] of the navigations and voyages of the sea and land made by Christopher Columbus, the excellent Genoese captain and valiant man, and others. The chart treats the islands and terra firma [i.e., mainland] explored up to the year 1535. With nautical signs, and complete knowledge of points of egress as well as places reached [the meaning is not entirely clear here].
Casatia, 9 January 1620
. . . Roberto Sanseverino

It is not entirely clear who this Sanseverino is; it is not the same figure as the Roberto Sanseverino (1418–1488), who married Elisabetta, daughter of Federico di Montefeltro, Duke of Urbino; he lived in the fifteenth century, not the seventeenth. Nor is it clear why this chart or map has been made, apparently over a century after Columbus's voyages. Finally, there is the question of why the legend says "explored up to the year 1535"—this date does not seem to have any particular significance in the history of voyages to the New World, at least as related to this document.[10]

In fact, the map bears some dotted lines, apparently indicating at least part of these explorations. But the map seems a composite of at least two maps: the islands of Cuba and Ispaniola are oriented as we might expect, with west to the left, and east to the right sides of the parchment. But the terra firma seems oriented with north to the left-hand side, thus putting Mexico to the right.

The toponyms listed on the parchment are as follows:

Mare Atlantico
Terra fiorita
Cuba
Ispaniola
B. Vergine

Nova Ispania
Messico Impero
peninsola
Selvana
Mare oppositu

We can decipher some of these toponyms, as we reorient the components of the map slightly. If we turn the main part of the map so that north is at the top, while retaining the Caribbean islands configured as they are, we can more clearly see what is being represented here (see fig. 16). Now we see that the Mare Atlantico is the Atlantic Ocean, to

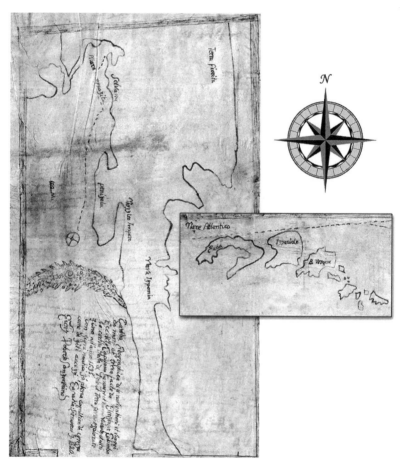

16   The "Columbus Map" with the Americas reoriented north to south.

the east. The southeast coast of North America labeled "Terra fiorita," that is, Florida. Florida was reached by the Spanish explorer Juan Ponce de León in the early sixteenth century. The toponym here is not unusual, and, in fact, Sebastian Munster's 1550 map of the New World has a label on the North American continent reading "Terra Florida." Ispaniola is the island of Hispaniola, where Columbus arrived on his first voyage, in 1492. The toponym B. Vergine might indicate Beata Vergine, and so perhaps indicates an island of the "Blessed Virgin." It is not entirely clear what the interpretation should be here; perhaps this is what is today Puerto Rico.

We see to the west Nova Ispania, that is, "New Spain," and above that Messico Impero, that is, the "Empire of Mexico," apparently referring to the Aztec empire, which was toppled by Cortez. Farther to the west, we see the simple label "peninsola," which we might read as the Baja peninsula. The territory above it is labeled only "selvana," that is, some kind of wilds or jungle. We see another dotted line to that area, oddly coming from a circular mark in the water off the aforementioned peninsula.

That body of water near this *selvana* region on the map is labeled "Mare oppositu," that is, "Opposite Sea"—the implication is that it is on the opposite side of the world from the Atlantic Ocean, or opposite in terms of being on the other side of Central America. Beyond the peninsula, we see written "100 M.," apparently indicating distance. Where the inscription "Mare oppositu" is written, we see the coast curving around to the northwest and then to the west, with some kind of land bridge.

How does this map connect with the voyages of Columbus? Who drew it and why? How, too, did it end up in the same collection as the other documents, which pertain to the travels of Marco Polo? Once again, we are left with many questions. The only connections we can see here are the Sanseverino name and the depictions of new lands across the oceans.

SEVEN

Conclusions and Future Directions

Do these many maps and texts truly connect back to the famed tradition of Marco Polo and his narrative? Are they later copies of earlier works that are genuinely descended from the family of Marco Polo? Or were these documents, perhaps, fabricated in the eighteenth or nineteenth century as cartographic curiosities designed to deceive?

One can only speculate at this point, but there are a number of helpful questions we can formulate: If these documents do have genuine connections to the Polo tradition, then there might be some other references to them in the historical record. If the documents in this collection are pure fabrications, we must ask: Who put together such an elaborate series of works, with all the complex connections between the different pieces, and the several languages involved? And what might have been the motivation? In the case of the Vinland Map, possible motives for forgery might have been political or financial, but in the case of the Rossi documents, motives would be rather more obscure.

The documents themselves provide some clues to their origins or sources, and some of those clues are highlighted here.

The Ptolemaic Connection

As noted earlier in this book, one clue to understanding these maps and their possible origin is their Ptolemaic content. Claudius Ptolemy (ca. AD 100–170) lived and worked in Hellenistic Egypt, writing in the fields of astronomy and geography. His *Geography* is a comprehensive look at the *oikoumene*—the inhabited world—as far as he (and his sources) knew and understood it in that period. The *Geography* includes general information on the science of mapmaking itself, as well as long lists of coordinates of geographic locations, from which one can draw a series of maps covering the known world. No actual maps by Ptolemy's hand survive; however, his coordinates were used again and again through the centuries to create maps. As new discoveries were made, particularly in the fifteenth and sixteenth centuries, alterations and additions were simply incorporated into this ancient cartographic framework. It was not until well into the late Renaissance that mapmakers finally began anew, discarding the old Ptolemaic coordinates in exchange for purely empirical cartography.[1]

Ptolemy's work, however, was quite sophisticated for its time; it incorporated the latest knowledge of the lands and seas, and utilized reports from travelers. The coordinates were based on a longitude-latitude grid.[2] Ptolemy's work entered the Arab world, where it was utilized in a variety of later geographic writings, such as those of al-Khwarizmi (ca. AD 780–850).

In terms of the documents under investigation here, it is important to note that the *Geography* did not appear in a Latin translation until the fifteenth century.[3] However, in this collection, we clearly see Ptolemaic characteristics in maps that are apparently from a period much earlier than that. In the "Fantina Polo Map 1" (see pl. 6), we have a gridded map signed by Fantina Polo with the year "1329." Another such configuration is found in the "Moreta Polo Map 1" (see pl. 8) and in the "Moreta Polo Map 2" (see fig. 4a and 4b). These documents, too, bear dates from the fourteenth century, well before the entry into Europe of Ptolemy's geographic work. In fact, it was not until around 1400 that a wealthy and highly educated Florentine named Palla Strozzi brought a manuscript of Ptolemy's *Geography* from Constantinople to Florence. That was translated into Latin a few years later by Jacopo d'Angelo, an Italian scholar of Greek and Latin.[4] This raises two fundamental questions: Did the authors of these maps have access to a copy of Ptolemy's *Geography* before the rest of Europe did? Or are the maps from a later date than they purport to be?

Especially if the latter is true, whoever executed them was familiar with Ptolemy's work. The Ptolemaic characteristics are clear in a number of places here. First, there is the grid system itself, although it does not seem to have been laid out according to Ptolemy's indications of distance between grid lines. Then there is the distorted depiction of the British Isles, something we find in maps rendered from the Ptolemaic coordinates. In the drawing on the verso of the "Moreta Polo Map 2" for example (see fig. 4b), we see the northern part of the British Isles slightly "rotated" to the northeast. This distinctly Ptolemaic characteristic that we see in the "Moreta Polo Map 2" appears, for example, in a 1490 map based on the Ptolemaic list of coordinates for that region (see fig. 17).

In this "Moreta Polo Map 2," we also find a number of Ptolemaic toponyms, such as Serica, Cattigara, and Campalu. Other Ptolemaic features include the shape of the West African coast, which curves away to the southwest (see fig. 18). In addition, there is the peculiar rendering of South Asia. In Ptolemy, India is not a peninsula at all, but is given a coastline that runs flat from east to west. Just below this Ptolemaic India is the great island of Taprobane (modern Sri Lanka). In many early geographic works, this island is given huge proportions, as we see in Ptolemy. An oddity in the rendering in the "Moreta Polo Map 2," verso, and in the "Moreta Polo Map 1," recto, is that this massive Taprobane seems to be incorporated into India Gangem (i.e., India below the Ganges) itself (see fig. 19). In the Polo maps, along with this incorporation of Taprobane into the Indian subcontinent, however, we also see *another* isle of Taprobane to the southeast. This duplication of Taprobane, in fact, was not unusual in early maps.

There are some further interesting connections, not yet fully understood perhaps, between the work of Ptolemy and Marco Polo's narrative. At one point in the text of Ptolemy's *Geography*, we read of "1,378 islands" that are "around Taprobane."[5] Now this is the exact figure that appears in the 1486 Latin edition of Marco Polo (the one used by Columbus).[6] That is the third edition of the version of Polo's text penned by the Dominican friar Francesco Pipino.[7] Did the *first* edition of the Pipino version—which was put together about 1315–1320 and thus predated Ptolemy's introduction into the West—also have this figure? If so, it is possible that early versions of the Polo narrative somehow took in bits of information from Ptolemy.

Intriguing Clues and a Chinese Connection

As pointed out earlier in this book, the odd toponyms Uan Scian, To Qiú, Ta Can, and Fusan that appear in the "Fantina Polo Map 1," the "Fantina Polo Map 2," and in the "Lorenzo Polo Chronicle," may be

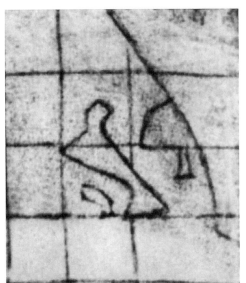

17   A comparison of the British Isles in the "Moreta Polo Map 2," verso (*at left*), and the British Isles in a 1490 edition of Ptolemy (*at right*).

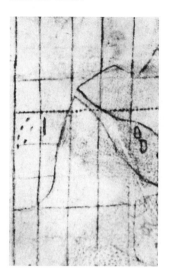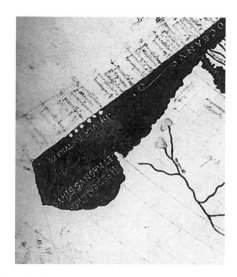

18   The West African coast below the Strait of Gibraltar in the "Moreta Polo Map 2," verso (*at left*) compared to the West African coast in an edition of Ptolemy (*at right*).

a vital clue to understanding at least part of this collection of materials. To a casual reader, these toponyms seem to have no meaning, but upon careful examination, we can see that they are Italian phonetic renderings of Chinese words. Moreover, they are derived from a very early Chinese legend, one concerning another toponym, Fusang (扶桑), a land across the ocean, allegedly very far from China.[8]

This legend is over a thousand years old. According to the narrative, a Buddhist monk named Hui Shen (慧深), originally from Afghanistan, and five other Buddhist monks are said to have traveled to lands in the most distant east, beyond China, during the reign of Emperor Ming in the Southern Sung Dynasty (AD 420–479). The narrative's details concerning the path by which Hui Shen reached this eastern land are not clear, and this ambiguity has led to a great deal of speculation as to where he and his companions actually went. Hui Shen called the land Fusang, a name that gives us few clues other than the fact that *sang* (桑) means "mulberry tree," and that the voyage there was by sea. The narrative provides some hazy details about Fusang, but nothing sufficient to make a positive identification. Various modern writers have claimed—and others denied—that the place discussed in the tale is, in fact, America. In the legend itself, of course, the descriptions are too vague to make any such assertion.[9]

What is important here is that this name, Fusang, appears in these

19 *Left to right*: India/Taprobane in "Moreta Polo Map 2," verso; India/Taprobane in document 8 ("Moreta Polo Map 1"), recto; and India/Taprobane in a typical Ptolemaic map, with encompassing outline added.

maps in the Rossi Collection, and that it is used to refer to part of the North American continent. This connection to North America, in fact, is repeated in later European cartographic interpretations of the Chinese tale.

The story of Hui Shen is found in several Chinese versions dating back to the seventh century AD.[10] However, the story did not reach Europe until the eighteenth century. Therefore, those who made the Polo maps that mention Fusang either were working after that time or drew the maps before that time, basing their depictions on the original Chinese texts or some early translations, and other source materials.

Moreover, we can actually identify some very clear matches between the odd toponyms in the Rossi maps and texts and the places mentioned in the story of Hui Shen. We can also find matches with locales discussed in other old Chinese accounts. In the version of the Hui Shen story found in the fourteenth-century *Wen xian tong kao* (文獻通考, "Comprehensive Studies in Literature"), we encounter a series of increasingly remote locales. First, we read of Wen Shen (文身); this term literally means "marked bodies" and may refer to a land inhabited by tattooed peoples, such as the Ainu. The text says that Wen Shen is situated to the northeast of Japan, but we are given nothing more specific than that. The text then speaks of a place called Da Han (大漢), a term that literally means "Great China." However, apparently this is not the land of China itself, since another Chinese text, the *Liang shu* (梁書, "History of the Liang Dynasty"), tells us that Da Han is found more than five thousand *li* east of Wen Shen, which itself is already beyond Japan.[11] After Da Han, the text tells us, is Fusang (扶桑).

We see a direct correspondence here to the toponyms found in the Rossi maps: Uan Scian = Wen Shen (文身), Ta Can = Da Han (大漢), and Fusan = Fusang (扶桑). We might also note that the story of Hui Shen says that beyond Fusang there is a "Kingdom of Women," much as the "Sirdomap Map" has an "Island of Women"; this is a place that also appears in the traditional Polo narrative.[12]

So, how did these Chinese toponyms come to appear on the Rossi maps? As noted above, the story of Hui Shen and the land of Fusang apparently did not reach Europe until the eighteenth century. In 1761, the French sinologist Joseph de Guignes (1759–1845), during the course of research on China, discovered and translated the accounts of Hui Shen's voyage.[13] There was much debate, however, as to how this land reported in the ancient Chinese texts actually should be interpreted. The cartographer Philippe Buache (1700–1773) put a land labeled "Fousang des Chinois" on a region of the Pacific Northwest coast of North

America on his 1752 map of these regions, the "Carte des terres nouvellement connues au nord de la Mer du Sud tant du Côté de l'Asie que de Côte de l'Amerique . . ." (see fig. 20a). Buache was working directly from the translation of de Guignes; the subtitle of this map was "Avec la route des Chinois en Amerique vers l'an 458 de J.C. tracée sur les connoissance geographiques que Mr. de Guignes a tirées des annales chinoises par Philippe Buache." On the map itself, we see the supposed route of Hui Shen, labeled "Route des Chinois en Amerique vers l'an 458 de J.C.," as well as the toponyms Venchin, Ta-Han, and finally Fousang—that is, Wen Shen (文身), Da Han (大漢), and Fusang (扶桑) (see fig. 20b). Antonio Zatta's map "Nuove scoperte de' russi al nord . . . ," which appeared in Venice in 1776, has the label "Fou-sang, Colonia de[i] Cinesi" in roughly the same locale as Buache's work.

In 1761, de Guignes presented a paper to the French Royal Academy on his findings concerning the account of Hui Shen. This paper was entitled "Recherches sur les navigations des Chinois du côté de l'Amérique, & sur quelques peuples situés à l'extrémité orientale de l'Asie."[14] In this work, he stated his belief that Fusang referred to Mexico. De Guignes went on to claim that the people and places described in the ancient Chinese account were the Indians of Mexico and the regions of the southwestern United States.[15]

It was not until many years later that these assertions were contested. In 1831, Heinrich Julius Klaproth, a German sinologist, attacked de Guignes's view.[16] But the debate was not over; Karl Friedrich Neumann, another sinologist, reiterated the original French interpretation and provided translations of the original Chinese texts. Charles Hippolyte de Paravey, also supporting the idea that Fusang referred to the Americas, generated two books on the subject.[17] In the United States, the discussions continued with Hubert Howe Bancroft, who treated the question of Fusang in his book *Native Races of the Pacific States*.[18] Another analysis by the sinologist Samuel Wells Williams appeared a short time later in a scholarly journal.[19]

However, the best-known work in America concerning this Chinese tale was that of Charles Godfrey Leland. Leland had been a student in Heidelberg, and there he had heard Neumann speak on the topic of Fusang. Leland sought to bring Neumann's ideas to America, and in 1850, they appeared in *Knickerbocker Magazine*. In 1875, Leland's treatment of Fusang came out in book form, with the publication of *Fusang; or, The Discovery of America by Chinese Buddhist Priests in the Fifth Century*.[20] This book provided Neumann's recounting of the story of Hui Shen, as well as a discussion of the navigation of the Pacific Ocean, and a look

20a Detail of the region of Fousang on Philippe Buache's 1752 map, the "Carte des terres nouvellement connues au nord de la Mer du Sud tant du Côté de l'Asie que de Côte de l'Amerique . . ."

20b Detail on Buache's 1752 map of the supposed route of Hui Shen, labeled "Route des Chinois en Amerique vers l'an 458 de J.C.," as well as the toponyms Venchin (= *Wen Shen* [文身]), Ta-Han (= *Da Han* [大漢]), and finally Fousang (= *Fusang* [扶桑])

at possible connections between American antiquities and Old World artifacts.

The fact that the Chinese toponyms Uan Scian, To Qiú, Ta Can, and Fusan appear on the Rossi maps may mean that these works are, then, from the eighteenth century or after. Alternatively, it could mean that these maps are of a much earlier date and are derived directly from the Chinese texts of the story of the travels of Hui Shen. In other words, the mapmaker was reading the Chinese text and taking the toponyms directly from the story and putting them onto his maps, transliterating the terms from the Chinese into Italian romanization.

If the mapmaker had simply made these maps from the later European translations of the Chinese tale of Fusang, then one might expect to find the toponyms put in the French romanization used by de Guignes, for example, Ven-chin and Ta-han. But instead, we find that the toponyms seem to have been rendered from the Chinese directly into an Italian transliteration: Uan Scian and Ta Can. For example, in English, the Chinese character 文 is rendered *wen*, whereas in Italian it would be rendered *uen*, since there is no "w" in that language. Similarly, the character 身 is rendered *shen* in English; de Guignes renders it as *chin* in French, with the "ch" serving to make the soft opening sound, and the "i" giving the appropriate short vowel sound. In Italian, the character would be transliterated as *scian*, to render this same soft "sh" sound—"ch" before "i" in Italian has a hard "k" sound, and so would not be appropriate. In Italian, the "c" before "i" has a soft sound like the English "child," and so the *sci* approximates the pronunciation of the Chinese character well.

De Guignes transliterates the Chinese characters 大漢 as Ta-han, whereas the Rossi maps have Ta Can; the use of Can to render 漢 may indicate an Italian dialectal artifact, where "c" is used to yield an "h" sound. The character 漢 is romanized in English as *han*. The evidence here, then, seems to indicate that the Chinese terms in the Rossi maps were being directly transliterated into Italian, and perhaps into the Venetian dialect in particular.

These are tantalizing connections—a series of obscure Chinese toponyms, rendered into Italian, with a possible Venetian author. We are faced with the fact that place-names that appear in a Chinese legend surface here in a set of maps and other documents related to the voyages of Marco Polo. If we conjecture that these materials are the product of modern fabricators, then these fabricators would have to have had the legend at their disposal.

But we must remember that the story of Fusang did not reach Europe until the eighteenth century. Again, it was not until 1761 that de Guignes translated from Chinese into French an account of the voyage of Hui Shen. That would mean that any possible fabricator of these Polo documents would have to have been working after this time. In addition, he would have had to take the francophone version of these Chinese place-names and "back-phoneticize" them into Italian, which seems unlikely.

The Context of the Documents

We have looked at the content of the documents and found an intriguing series of clues: (a) how the maps do—and do not—look like other early maps of the late medieval and early Renaissance period; (b) the fact that some of the documents contain Ptolemaic toponyms and other characteristics, such as a grid; (c) the several mentions of the famed if mysterious locale of Antilla; and (d) perhaps most curious of all, the presence of romanized Chinese toponyms on several of the documents.

In addition to content, however, we must also examine what we might term the "context" of the various pieces of the collection—their ownership, history, and so on. In the beginning of this book, a brief look was given to the apparent provenance of these materials, at least back to the owner who first showed them to the Library of Congress, Marcian Rossi. But what about ownership before that?

We are given a few clues by the documents themselves. The "Fantina Polo Map 2" has a small, oval-shaped tab attached to it at the bottom; the name on that tab reads "diana bonacolsin da Verona." The last name was a dialectical variation of the Italian name Bonacolsi. This was a famous Italian family that controlled several northern cities, including Mantua and Modena in the early thirteenth century. The "Sirdomap Map" also has one of these attached, oval-shaped tabs, on the verso. On this tab, we find a different name: Marta Veniero da Padova. The "Moreta Polo Map 1" has a tab with this same name. The name Venier is a Venetian form of the name Venerio. So we have a Venetian connection—indeed, Venerio was the surname of one of Venice's noble families. At this point, we can only speculate, but it is interesting that the maps in the Rossi Collection possess what seem to be tags of ownership. Bagrow believed that this Marta Veniero was the recipient of the map, sent by Moreta Polo, but this interpretation must also be seen as purely speculative.[21]

We find on the verso of the "Fantina Polo Map 2" the words "Tabula Geographica Fascio 96. Folio 254," and on the verso of the "Fantina Polo Map 1" the words "Tabula Geographica Fascio 96. Folio 255." The third example of this labeling is on the verso of the "Moreta Polo Map 1," where we read "Tabula Geographica Fascio 96. Folio 256." As noted in chapter 4, these labels may indicate that the documents were owned as part of a collection and suggest that we may find some as-yet-undiscovered related documents.

We know that Marcian Rossi possessed these documents, but we do not know much about prior ownership, other than what he claimed in his letter, discussed at the very beginning of this book. In that letter, he says that the documents came down through his family, the Rossi clan, to a "Marciano de Rossi," who Marcian Rossi said was his great-grandfather. We know that in 1887, Marcian Rossi immigrated to the United States from Italy; at that time, he was still a teenager. Some time after his arrival in the United States and prior to 1906, it seems that he returned to Italy to collect the documents that he had inherited.

It becomes difficult at this point in the investigation to know what Marcian Rossi possessed, since some pieces of the collection seem to have been lost, sold, or traded away. Our earliest clue in terms of his possession of these materials is from 1904; in that year, a map and a letter of a certain "Taddeo Visco of Geneo" (whose name appears elsewhere in the collection, as noted above) were exhibited at the anthropology section of the St. Louis Exposition. This was reported in a French journal in a few—rather disparaging and anonymously written—lines: "[L]a section d'Anthropologie avait réuni, au point de vue de l'historie coloniale et celle de la Louisiane en particulier, des documents de tout premier ordre. A côté d'eux, quelques pièces très contestables. On ne voit pas trop ce que peut représenter la carte de Taddeo Visco de Gênes; on doute à bon droit de son authenticité, et la lettre qui l'accompagne paraît également suspecté."[22] Reproductions of these "Taddeo Visco" materials are in a file at the Geography and Map Division of the Library of Congress, but the originals are missing.

Despite the mention in the French journal that the "Taddeo Visco" material might be "suspecté," evidence from this period points to continued serious interest in the documents possessed by Rossi. In an archive in Paris, there are three letters from a Frenchman surnamed Eichard.[23] The first letter is dated 1 February 1906, and is addressed to

"Monsieur le Baron Hulot, Secrétaire Général de la Société de Géographie" in Paris.[24] Eichard is writing from San Francisco, and it is clear in the letter that he has met Marcian Rossi. Eichard reports that he is sending four photographs of documents that belong to Rossi. In the letter, Eichard mentions Rossi's mother, referring to her as "D. Cristina Rossi à Caserta."

The second letter is of the same date and is also addressed to the same person. In this letter, Eichard describes the contents of the above-mentioned photographs—images of the "Taddeo Visco" documents. The letter goes on to say that one of these documents bears the signatures of the father of Christopher Colombus and of Christopher Colombus himself, and that it was found "dans la bibliothèque de l'oncle de M. Rossi, Monsieur Giacomo Cavicchia à Latina, province de Terra di Lavoro en 1884." Eichard provides a translation, in French, of one of the documents, a "lettre sur parchemin de Taddeo Visco"; that text recounts a very peculiar story of how a Roman expedition under Pliny the Elder came to the New World (specifically, the Antilles) and discovered an Etruscan colony there. No other historical source matches this odd tale, however, and it is not found in the extant work of Pliny.

The third letter is dated 15 October 1906 and again is addressed to the "Secrétaire Général de la Société de Géographie," but this time it seems that Marcian Rossi is the writer, although the letter is in French, not Italian or English. In this letter, Rossi makes further comments on the matter discussed above, elaborating that in Italy there are actual "parchments" written by Pliny the Elder describing his voyage to the Antilles, and the finding of Etruscans there. Rossi also states that at the 1904 St. Louis World's Fair, he displayed this manuscript, along with a map of the Antilles made by Pliny, and a letter from Taddeo Visco to the Marchese di Sessa (i.e., the Marquis of Sessa) dated 1439, talking about this same story. Rossi then mentions an article in the *St. Louis Republic* newspaper that discusses these documents. Finally, Rossi notes that all the original materials were lost in the San Francisco fire (which took place in April of 1906), and he requests copies of the photographs that Eichard had made. Unfortunately, these photographs are now missing.

The trail takes another turn, with a mention in 1920 in, of all places, a Caracas newspaper, *El Nuevo Diario*. On 6 August of that year, an article appeared, written by Alberto Francisco Porta, who worked as both an architect and a professor at Santa Clara University. In fact, as noted in the introduction to this book, Porta was a friend of Marcian Rossi. The article mentions that Rossi is in possession of the "Taddeo Visco" documents—or at least reproductions of them, since we recall that he claimed that the originals were lost in the San Francisco fire. The reproductions, then, must have disappeared—perhaps sold or traded away—after this time. Professor Porta himself was an interesting figure; he taught architecture and delved into the fields of engineering, history, archaeology, and even astronomy. It is also not surprising that he was friends with Rossi—both were from Italy, both were interested in history, and perhaps most important, both were intrigued with the idea of pre-Columbian voyages to the New World.

In 1929, there was the following entry in the academic journal *Italica*:

The University [of California, Berkeley] has now on exhibition many old manuscripts, discovered by Professor Altrocchi at San José and belonging to Mr. Marciano F. Rossi. They include mediaeval accounts, one of them dated 1295 and containing the name of Marco Polo, papal bulls and documents, an extraordinary description of a trip by an Italian along the coasts of California in the early XVIth century, and a MS volume given by Clement VIII to Torquato Tasso, with probably autograph annotations by the latter (which Professor Altrocchi is studying).[25]

This is Rudolph Altrocchi, professor of Italian at University of Chicago from 1915 to 1926, then at Brown in 1927, and at Berkeley after 1928. Obviously, he believed that the Rossi documents merited serious study. Apparently, though, Altrocchi never published on the subject. It seems, moreover, that some of the materials that are mentioned here as belonging to Rossi are now missing, for example, the "description of a trip by an Italian along the coasts of California in the early XVIth century," and the "MS volume given by Clement VIII to Torquato Tasso."

The next mention of the maps occurs in 1933. This is around the time Marcian Rossi contacted the Library of Congress. In 1933, a series of articles on the "Marco Polo" materials appeared in a number of different newspapers, all on the same day. The headlines ran "Map of Marco Polo Trip Was Made by Daughter" (*New York Times*, 28 December 1933), "Marco Polo Map of 13th Century Believed Found" (*Chicago Daily Tribune*, 28 December 1933), and "Marco Polo Map Shown to Experts: Copy of Californian's Chart Indicates Origin in Fourteenth Century" (*Los Angeles Times*, 28 December 1933). A few days later, there was a story entitled "Marco Polo Map Found" (*Minneapolis Journal*, 1 January 1934). In 1936, the *Christian Science Monitor* featured an article entitled, "Marco Polo's Log Comes to Light."

Studies and Investigations

Despite this flurry of popular interest, still no academic study appeared. In 1943, Marcian Rossi gave the "Map with Ship" to the Library of Congress—in his continued efforts, it seems, to arrange a serious examination of the collection. It is important to note that none of the correspondence between Rossi and the Library of Congress indicates any desire for monetary reward or even publicity for himself. Rossi's interests appear to have been genuine and reflecting an honest desire for inquiry. Sadly, Rossi passed away in 1948, and it is only in that year that a serious if incomplete study of the maps appeared—Leo Bagrow's article in the journal *Imago Mundi*.

Despite the obvious attention the Rossi materials received in the period prior to Rossi's death, after that time the documents in some sense fell into obscurity. As noted at the beginning of this book, some time in the 1930s a brief investigation was done by William J. Wilson of the Library of Congress, although a more complete study appeared a couple of decades later.[26]

In 1949, Louis Arthur Rossi (1898–1990)—Marcian's son—took some interest in the maps, commencing correspondence with the Library of Congress. But it is not until 1968 that a dated receipt from February of that year comes to light, indicating that Melvin H. Jackson, chief of the Division of Maritime History at the Smithsonian Institution, had received eleven documents for the "purpose of examination and evaluation" from Louis Rossi. A memorandum dated 11 March 1968 from Walter W. Ristow, chief of the Geography and Map Division at the Library of Congress, to the director of the Reference Department, notes that Jackson has the Rossi documents for "study and examination." Still, no formal paper was published, nor is it clear if such an "examination and evaluation" or "study and examination" were ever actually carried out.

Unlike Marcian, it seems that Louis Rossi was indeed interested in the financial potential of the documents. We find a letter dated 19 June 1979, from Juan G. Frivaldo,[27] business partner of Louis Rossi, to a "Mr. Hossein," seeking to sell the documents. This letter includes Frivaldo's statement that he has Rossi's authorization to carry out such a sale. In a letter dated 10 August 1979, Louis Rossi wrote to a "Mr. Eddi Roberto" stating, "This will authorize you to offer for sale 9 Parchment Maps of Marco Polo's visit to China and other historical Parchments [*sic*]." Finally, in a memo dated 3 January 1980, Louis Rossi states that he has "authorized Gov. Juan G. Frivaldo of 161 W. Santa Clara St., San Jose, California 95113 as my sole Literary Agent with specific duty to . . . sell rare sheepskin manuscripts believed written by the Family of the famous Venetian traveler Marcos Polo [*sic*]."

However, it appears that by the mid-1980s Louis Rossi had abandoned his attempts to sell the documents. In 1990, when Louis Rossi passed away, the documents became the property of his daughter, Beverly Pendergraft (née Rossi), with actual possession of the materials finally going to her son, Jeffrey R. Pendergraft, the present owner.

Throughout this whole period, a complete investigation was never carried out, and therefore we arrive at the present time with an intriguing set of documents surrounded by a great deal of uncertainty as to their precise provenance (at least in the centuries prior to Marcian Rossi), and a fair measure of obscurity as to their actual relation to the voyages and narrative of Marco Polo. The key question, of course, is Are these maps what they purport to be?

How Do the Documents "Fit"?

The famous Vinland Map in some ways set the stage for a certain kind of cartographic controversy about authenticity. Some still hold that the Vinland Map is what it purports to be. In the case of this collection of "Marco Polo Maps," we must proceed with caution, since there are so many questions raised by these documents. In the case of the Vinland Map, there is just one document. Moreover, in that case, the motive for forgery is quite clear: to provide cartographic "evidence" of Viking explorations of the New World. In the case of these "Marco Polo" documents, the motivations of a possible forger are less clear. To prove that Marco Polo ventured toward the New World? To indicate Chinese or Arab involvement in voyages to the New World? It would have been much simpler to fabricate a map with toponyms from the Polo narrative, and simply add a portion of the New World. There would have been no need to include the complex and confusing details from Ptolemy and Pliny, the Chinese characters, and so on.

Perhaps the documents, if fabrications, were made for financial gain, but we have no indication that Marcian Rossi wanted to *sell* them to the Library of Congress—rather, he simply seems to have been interested in finding out more about them. Moreover, if these documents are fabrications, when were they created and by whom? And why are there so many? There are some fourteen documents in all, with both figures and text; the documents include passages in Italian, Latin, Chinese, and Arabic. The content of each document connects in some way to the other documents, revealing a fairly complex narrative. A fabricator would have to be quite knowledgeable—and dedicated—to create such a large body of work.

It is also interesting to note that the Sanseverino clan, the della Rovere clan, and to a lesser extent the Rossi family were all members of the Italian nobility and related by marriage to one another and to other noble families in Europe. This in and of itself does not prove anything, but it may offer some explanation as to how and why such important documents could have been preserved and passed down over a period of several hundred years.

Certainly, the handwriting on several of the documents seems to suggest dates later than those given in the text of the materials themselves; a full paleographic study is required. The handwriting as it appears on the documents could mean that as they stand now they are fabrications, or modernized copies of documents that really do date back to the time of Marco Polo's daughters, that is, the fourteenth century. It is also could be that the true nature of this collection reflects some kind of combination of these two possibilities.

Bagrow, in his preliminary investigation of the documents, felt that the documents were "more or less late modernized copies," and added that the "copyists have not endeavored to conceal the time at which the copies were made. In their script there is no effort to imitate the characteristic oldness of each original."[28] This may be the case, but the collection may actually present an even more complex scenario, one which will be elucidated only with further investigation. Such an investigation would have to answer the question of who these "copyists" were, what the original materials might have looked like, when the copying took place, and so on.

If we posit the idea that these documents were simply fabricated and are not what they purport to be, again we have the question of who made them. As pointed out earlier, the documents are not simple creations—they involve multiple languages, personages, cartographic renderings, and so on. There are also many particulars in the documents, details that would have required reasonably sound cartographic knowledge on the part of the fabricator. These particulars include the mention of the famed mythical island of Antilia, the Ptolemaic nature of some of the toponyms, and the presence of Arabic and Chinese writing on some of the documents.

Part of the challenge here is that the Rossi maps in some ways both stand apart from extant historical sources and are closely tied to those sources. These "Marco Polo Maps" seem unique in their depictions but at the same time have toponyms and characteristics that find echoes in other cartographic and historical materials—toponyms and characteristics such as Antilla, Ptolemaic cartography, and so on. These perplexing observations are compounded by the fact that we have very little information on Marco Polo or his family. As one writer has commented:

Apart from interpretation and extrapolation, very little is known about Marco Polo save the bare account in the Prologue to the *Description of the World*, which simply tells us that he lived abroad for twenty-six years and that, in 1298, whilst in prison in Genoa, he wrote down his account of those years. Beyond this simple outline of how the years between 1271 and 1295 were spent, what little information we have about Marco Polo and his family comes from other, secondary sources. There are references in Jacopo da Acqui's *Imago Mundi*; a small number of surviving documents in the Venetian archives regarding minor legal battles; and the longest account, unfortunately not entirely reliable, in Giovanni Battista Ramusio's *Navigationi et Viaggi*, published in 1559.[29]

The documents in this collection bear the names of a number of figures, including "Guido Spinola," "Carlo Sperano," "Lorenzo Polo," "Bellela Polo," "Fantina Polo," "Moreta Polo," and others. Marco Polo dictated his last will and testament on 9 January 1323, with a priest and notary present. That documents reveals that at this time, Marco's wife Donata was still alive, as were their daughters, Bellela, Fantina, and Moreta.[30] Unfortunately, however, the will of Marco Polo does not mention maps and texts, nor any of the persons named above.

If one is to speculate that the Rossi maps are pure fabrications, then possible suspects must be sought. One can only put forward some names, but there is little in the way of evidence or even motive. Some years ago, a scholar suggested that the Vinland Map was forged by the Jesuit cartographer Father Josef Fischer (1858–1944), with motives that were political.[31] It would be interesting to speculate that Fischer was also responsible for the Rossi maps, but there seems no apparent connection.

For the Rossi maps, it might be that the sinologist Joseph de Guignes was responsible, but why would he have created maps that suggest distant travels beyond China by Marco Polo? De Guignes's concern was demonstrating that the Chinese—not the Italians—had come to the shores of America in ancient times. It is interesting to note that even earlier than de Guignes, the Portuguese writer António Galvão, in his

1563 work *Tratado dos descobrimentos*, had claimed that the Chinese had reached the New World, including Central America, South America, and the Caribbean.[32] But there seems no direct connection between Galvão's work and the maps in this collection.

Since the text in the Rossi documents is primarily in Italian, it might be suggested that an Italian sinologist was responsible. We have Father Martino Martini (1614–1661), a Jesuit missionary to the Chinese, who worked in the areas of theology, history, and cartography. He produced the extensive *Novus Atlas Sinensis*, a series of maps of all of China. There is, too, the other famous missionary Matteo Ricci, who also worked extensively in the area of cartography while in China. But we can only speculate; the standard histories certainly contain no account of an Italian sinologist fabricating maps of this kind.

The rather extensive literature concerning the mythical "Straits of Anian" could provide a clue, since these Rossi documents suggest that Marco Polo voyaged to the ends of Asia, across a strait filled with a chain of islands, to a mainland beyond.[33] Concerning such a navigable passage, there are certainly well-known fakes and frauds in the historical record. Lorenzo Ferrer Maldonado, a sixteenth-century Spanish navigator, wrote a work entitled *Relación del descubrimiento del Estrecho de Anian*, describing a voyage that was probably fanciful.[34] The French geographer Buache obtained a copy of this work and put a strait between Asia and North America—along with Fousang as a region on the Pacific Northwest coast of North America—on his maps, as noted earlier.[35]

The Polo Voyages and Maps, Known and Unknown

There is no doubt that the narrative concerning Marco Polo's voyages influenced a great deal of subsequent mapmaking. Toponyms from the narrative appeared on maps even centuries after those travels, in fact.[36] But the historian John Larner carefully dissected the cartographic influence of the narrative of Marco Polo (what Larner calls "the Book"), and he points out that such influence is not at all clear. Larner notes that of the several hundred world maps that survive from the fifteenth century period (prior to the 1492 voyage of Columbus), most are either traditional medieval maps or works based on Ptolemy's coordinates. Significant mapmakers of this period, such as Andreas Walsperger, either "ignore Marco's Book" or use information from the Polo narrative in a "very limited and superficial" way.[37] Larner says that it was not that the cartographers disbelieved the Polo account, but rather that they were overwhelmed by the number of sources that they had to assess and utilize in their maps, sources that included—besides Marco Polo—Ptolemy, classical sources, Christian works, and so on.

It is only with the world map of Fra Mauro that we see toponyms from the Polo narrative utilized in great measure, throughout Mauro's rendering of China, and with the inclusion of the "ixola de Zimpagu," that is, Marco Polo's "Japan."[38] Chronologically from there, Marco's toponyms are used in the "Yale" map from circa 1489 by Henricus Martellus, and the 1492 globe of Martin Behaim. In fact, Behaim seems to have used at least two different translations of the Polo narrative, one in Latin and one in Italian.[39]

While the maps in the Rossi Collection suggest the possession by Marco Polo of knowledge of fragments of the New World—for example, coastlines on the other side of the Pacific Ocean—the traditional connection between the Polo narrative and lands across the sea is rather different. Well known, of course, is the fact that Columbus had read about the travels of Marco Polo before undertaking his transatlantic crossings, and that he seems to have believed at times that he had reached Asia on the other side of the Atlantic Ocean.[40] At one point, one of Columbus's captains even "suggested that they . . . steer to the southwest in search of the great island of Çipango."[41]

But while we can see the influence of Polo's narrative, there is nothing to suggest that Columbus was aware of any actual *map* from the time of Polo or Polo's daughters. Indeed, several documents here in the collection—the "Moreta Polo Map 1," the "Moreta Polo Map 2," and the "Map with Ship"—suggest a distinct *separation* between the farthest reaches of Asia and the New World, while Columbus seems to have felt, at least initially, that his discoveries were not really of a New World but rather that they had revealed the islands and coasts of East Asia.

Of course, we should ask whether there ever existed any maps at all that *directly* derived from Marco Polo's voyages. We find some traces; as Yule notes, "We are told that Prince Pedro of Portugal [1392–1449] in 1426 received from the Signory of Venice a map which was supposed to be either an original or a copy of one by Marco Polo's own hand."[42] With a rather trepidatious double negative, Yule gives his own view: "There is no evidence to justify any absolute expression of disbelief; and if any map-maker with the spirit of the author of the Carta Catalana [a famous fourteenth-century cartographic work] then dwelt in Venice, Polo certainly could not have gone to his grave uncatechised. But I should suspect the map to have been a copy of the old one that existed in the Sala dello Scudo of the Ducal Palace."[43] As we will see, this echoes a comment by Giacinto Placido Zurla, a learned cardinal vicar of Rome, born in 1769, about a map in the Ducal Palace (the "Palazzo Ducale") in Venice. Yule continues: "The maps now to be seen painted on the walls of that Hall, on which Polo's route is marked, are not of any great interest. But in the middle of the 15th century there was an old *Descriptio Orbis sive Mappamundus* in the Hall, and when the apartment was renewed in 1459 a decree of the Senate ordered that such a map should be repainted on the new walls. This also perished by a fire in 1483."[44] The direct visual evidence, unfortunately, is lost, then.

We have, however, discussions and descriptions. Ramusio's 1558 *Navigationi et Viaggi* speaks of a map brought back from Cathay by Marco Polo. The passage is worth quoting in full:

About this book, it remains for me to say many things in general that I heard many times when I was young from the most learned and reverend Don Paolo Orlandino of Florence, an excellent cosmographer and a good friend of mine. He was the prior of the Monastery of San Michele di Murano, near Venice, of the Camoldese Order. He told me that he had heard these things from other old friars in his monastery—that the beautiful, antique map with miniatures on parchment, which can still be seen today in a large armoire alongside the choir of their monastery, was for the first time diligently copied and drawn by one of the lay brothers of the monastery who took delight in the study of cosmography. It was diligently drawn and copied from a very beautiful and very old marine chart [*carta marina*] and from a world map [*mappamondo*] that had been brought from Cathay by the great Sir Marco Polo and his father. He [i.e., Marco Polo], as he travelled through provinces under order of the Khan, kept adding and annotating on his maps the cities and locations that he found there, as is described above.

However, because of the ignorance of another man who rendered and produced it after him, adding descriptions of men and animals of different types and other absurdities, many modern and rather ridiculous things were added, so that for many years it lost all authority with men of judgment.

But, since, not many years ago, judicious people began to read and consider this book by Sir Marco Polo more diligently than they had before, and to compare what he writes with the picture [of this map], it immediately became clear that the aforementioned map was without any doubt derived from that of Sir Marco Polo. [The map] was from the beginning based on that [book], with accurate measurements and in good order.

For this reason, right up to today, it has been held in such veneration and prestige throughout this city, and, by those who delight in the things [concerning] cosmography, such that there is not one day when this book would not be seen and considered with great pleasure by some of them. And among the other miracles of this divine city, when foreigners go to see the works in glass in Murano, [this book] might [also] be shown as a beautiful and rare thing.

And, although in the map many things are done rather confusedly and without order, [such as] degree and measurements (something that ought to be attributed to the person who rendered and produced it), there are included many beautiful and worthy details, not yet known even by the ancients—such as, towards the Antarctic, where Ptolemy and all other cosmographers place *terra incognita* without the sea, in this map from San Michele di Murano, made so many years ago, one see that the sea encircles Africa and that one can sail there heading westward. Such things were known at the time of Sir Marco, although no name was yet given to the cape that [came to be] called "Good Hope" by the Portuguese in our times, in the year 1500.[45]

Unfortunately, the passage does not provide us as many details as we might like concerning what this map looked like.[46] However, it is interesting to note that the Monastery of San Michele di Murano mentioned here was the monastery of Fra Mauro—and it seems that his map is being discussed here.

The text above speaks of Polo "adding and annotating on his maps the cities and locations that he found there"—but the Rossi maps we have presented in this book are, in fact, rather devoid of the toponyms one finds in the traditional narrative. Certainly, the renderings found in the Rossi Collection include both smaller "regional maps" (if not a true *carta marina*) and world maps, but they seem, alas, unconnected to this discussion in Ramusio. As one commentator puts it, Ramusio believed that the "original maps brought to Venice by Marco Polo and his father had become corrupted" and that "Marco Polo's (presumed) original cartographic documents had been already lost around the mid-fifteenth century."[47] In other words, "Ramusio takes for granted that Marco Polo must have prepared the geographic maps and that these might have been transmitted, more or less corrupt, until Fra Mauro's times."[48]

This leads to two points to consider. The first is that Ramusio seems to be completely in error, and that the Polos, in fact, brought back no maps whatsoever. The second is that any presumed fabricator of these "Marco Polo" works made no effort to utilize this material from Ramusio or bother to use the Polo narrative.

The last paragraph of the Ramusio passage, though, is interesting. First, the reader should be impressed by Ramusio's honesty, where he fully admits that the map, though admirable, includes "many things . . .

done rather confusedly and without order, [such as] degree and measurements." Also, we see reflected a very typical medieval and Renaissance idea that the ancient world actually possessed *more* knowledge than those in the present time—Ramusio remarks with surprise here that the map has details "not yet known even by the ancients."[49]

Most interesting, in terms of cartography, is what Ramusio says a few lines after the passage just cited: "[T]oward the Antarctic, where Ptolemy and all other cosmographers place *terra incognita* without the sea, in this [map] of San Michele made already many years ago, one sees that the sea encircles Africa and that one can sail there toward the west. [This] was known at the time of Sir Marco, although no name was yet given to the cape that [came to be] called 'Good Hope' by the Portuguese in our times, in the year 1500."[50] Various cartographers had pondered—prior to the rounding of the continent by Vasco da Gama (1460–1524)—that Africa might be surrounded by ocean to the south. But Ptolemy's map did not show this, as Ramusio correctly remarks. For a map to show Africa as a peninsula prior to the Portuguese discovery is, however, not that surprising: speculations to this effect date back to the classical period.[51] More pertinent to our discussion here, however, is the fact that the "Marco Polo Maps" again do *not* seem to resonate with the description in Ramusio, as they barely show Africa at all. In other words, what Ramusio says "was known at the time of Sir Marco" is *not*, in fact, reflected in our "Marco Polo Maps" at all.

The internal evidence of the Rossi documents imply that, as Ramusio suggests, Marco Polo brought cartographic materials back with him upon his return. But other evidence leads us to doubt that Polo brought back maps—Chinese or otherwise—from Cathay.[52]

Some Final Hints and Future Directions for Research

However, another clue comes from Zurla. Zurla was a theologian and historian and was particularly interested in the history of cartography. In 1806, he published a study concerning the work of Fra Mauro. Later, he wrote a two-volume work entitled *Di Marco Polo e degli altri viaggiatori veneziani*.[53] Soon after that, a shorter cartographic study by Zurla was published in Venice, entitled *Sulle antiche mappe idro-geografiche lavorate in Venezia*.[54]

Part of that latter work examined a map of the voyages of Marco Polo—a map in the Palazzo Ducale (known as the Ducal Palace or Doge's Palace in English) in Venice. This is an Italian palace in the

Gothic style that was built in various stages from the fourteenth to early fifteenth century. This was the residence of the *doge*—that is, the duke—and contained his offices and those of other officials. In the palace, there is a map room, the Sala dello Scudo. There, one finds a large world map, oriented with south at the top. What Zurla says in his book concerning this map is quite interesting. First he describes the map, and all the regions it shows; he then notes that parts of the map's depictions are like the descriptions found in Ramusio.[55]

More specifically, Zurla examines the portion of the map that shows North America. He argues that the northwestern portion of the continent was there on an *earlier* map—a map that served as the source for this one. He comments that "this primitive indication of a large island or continent in the old maps is most interesting."[56] Zurla goes on to imply that Marco Polo somehow had knowledge of these regions, including the strait between Asia and America, "cinque secoli"—five centuries—before the voyages of Bering and others.

As far as we know, "Marco Polo drew . . . no maps recording his experience."[57] Maps such as the 1375 Catalan Atlas use information from the Polo narrative in the depiction of Asia, but there is no suggestion that this or similar cartographic works were based on any maps that Marco Polo or his fellow travelers might have brought back. Fra Mauro's world map uses Marco Polo's toponyms, and later Ramusio claimed that this work was a copy of one brought back from China by Marco Polo. However, there is no evidence provided for this claim.

Zurla's comment about a "primitive indication of a large island or continent in the old maps," though, is indeed intriguing.

In terms of the documents themselves, some testing could provide further insights into their authenticity. Radiocarbon dating can be done on the parchments; however, even if the parchments are genuinely centuries old, this does not mean that the maps drawn on them are from the same time period—old vellum can be erased and reused.[58] Therefore, studies of the ink could also be carried out; these are typically used to determine the composition of the ink in a given manuscript. To simplify matters somewhat, if the ink is found to contain modern ingredients, then, of course, the work is considered probably a forgery.[59] In addition, as noted earlier, a complete paleographic study should be done.

What *are* these documents? It is possible that they are fabrications, but that conjecture presents many questions. The collection includes a number of documents, which in turn are closely inter-related in terms of content. Moreover, there is no clear answer to the issue of cui bono—

there seems, for example, to have been no financial motive here, including on the part of Rossi himself. Particularly puzzling, as discussed earlier, is the presence in the documents of the place-names from the Chinese story of Fusang —it seems that whoever penned those would have had to know of the story before it was brought to Europe by de Guignes in the eighteenth century.

Earlier, another possibility was presented: that the collection might represent a combination of two possibilities, that is, that the collection includes fabrications and modernized copies. If some of these materials are modernized copies of documents that, in fact, date back to the time of Marco Polo and his daughters, it would be worthwhile determining what elements have been modernized and by whom. Regardless, in that case, we would have something truly important here—evidence that there is more to Marco Polo than his *Il Milione* narrative. We would have evidence, then, that Marco Polo and his daughters founded a separate cartographic tradition concerning travels to the distant east, a tradition of which only these traces remain.

Acknowledgments

Limitations of space prevent me from including all the names of the many individuals who assisted me in the research and writing of this book. However, I especially wish to thank Jeffrey R. Pendergraft, for his allowing me complete access to the maps and other materials in his family's possession, and his open and unbiased views about the research. I also wish to thank Richard and Victor Rossi, who provided very valuable information about the Rossi family genealogy. Further assistance in this matter was kindly given by Pier Luigi Poldi Allaj of the Associazione Culturale Corte dei Rossi in Italy.

My investigations into the maps were assisted by John R. Hébert and Ron Grim of the Library of Congress, and I also appreciate the thoughtful input of John Hessler there. Vladimir Valerio was extremely helpful with his comments about the Rossi Collection generally, and in his interpretations and transcriptions of the Italian text on the maps. I also greatly appreciate the Society for the History of Discoveries, Connie Brown and Fred Schauger of the New York Map Society, and Ernst "Ernie" Hamm of York University for inviting me to give lectures on these maps.

Carla Weinberg and Rosella Diliberto provided assistance with some of the Italian text, and I extend many thanks to Joe Farrell of the University of Pennsylvania and Marc Cohen for their input on the Latin passages. Nathan Sivin offered suggestions concerning the Chinese characters that appear on some of the maps. I also wish to thank Carol Graney, Sara J. MacDonald, and Mary Louise Castaldi of the University of the Arts Libraries for their consistent help in gathering often obscure secondary sources. The University of the Arts provided a grant that helped me to conduct some of the background research for this book.

On a personal note, my wife Charlotte Lin provided encouragement and moral support during the long years of research required by this project.

I also wish to thank Erik Carlson, Nicholas Lilly, and the other staff at the University of Chicago Press for their diligent work in carrying this project to completion. Abby Collier, formerly of the University of Chicago Press, was a wonderfully patient, attentive, and meticulous editor, and any mistakes in this volume are most certainly mine, not hers.

Appendix 1: An Inventory of the Documents

As noted in the text, there are fourteen documents in the Rossi Collection that discuss Marco Polo and his travels. Below is an inventory, with the current numbering, titles, descriptions, and list of references. The abbreviations for those references are as follows:

Bagrow = Leo Bagrow, "The Maps from the Home Archives of the Descendants of a Friend of Marco Polo," *Imago Mundi* 5 (1948): 3–13.

Black = John Black, "Marco Polo Documents Incorporated in the Felicitation Volumes of Southeast Asian Studies," in *Felicitation Volumes of Southeast-Asian Studies Presented to His Highness Prince Dhaninivat Kromamun Bidyalabh Bridhyakorn on the Occasion of His Eightieth Birthday*, ed. Prince Dhaninivat Sonakul, 2 vols. (Bangkok: Siam Society, 1965).

LAR = Louis Arthur Rossi (1898–1990); this is a personal inventory originally assembled by Louis Arthur Rossi, currently held by Jeffery R. Pendergraft.

Document Number: 1
Title: "Sirdomap Map"
Description: A map of northeastern Asia with toponyms, referred to as the "Sirdomap Map," with Arabic lettering; the map has a short text in Italian below it, apparently written by Bellela Polo, and mentioning a certain Syrian navigator named "Biaxio Sirdomap." The map has toponyms referred to by a series of Roman

numerals. See the descriptions below of document 2 ("Sirdomap Text") and document 3 ("Bellela Polo Chronicle").
References: Bagrow, 6–9 and fig. 6; LAR, 6.

Document Number: 2
Title: "Sirdomap Text"
Description: A short text in "Arabic" lettering, followed by a brief Italian text, with the year "1267." Allegedly, this text was written by Sirdomap; see document 1 ("Sirdomap Map") and document 3 ("Bellela Polo Chronicle").
References: Bagrow, 6–10; LAR, 7.

Document Number: 3
Title: "Bellela Polo Chronicle"
Description: A text concerning Marco Polo and Sirdomap; this text is in Italian, and apparently written by Bellela Polo. See document 1 ("Sirdomap Map") and document 2 ("Sirdomap Text").
References: Bagrow, 6–9, LAR, 9.

Document Number: 4
Title: "Map with Ship"
Description: A map of eastern Asia, along with a picture of a sailing vessel; it is currently held by the Library of Congress. This map has inscriptions in Italian, Arabic, and Chinese.
References: Bagrow, 5–6 and fig. 5.

Document Number: 5
Title: "Pantect Map"
Description: A map of eastern Asia, with an attached accompanying text, referred to as the "keynote" by Bagrow. This "keynote" is now missing; see the description of document 13 ("Keynote to Pantect Map").
References: Bagrow, 5 and fig. 4; LAR, 3.

Document Number: 6
Title: "Fantina Polo Map 1"
Description: A map covering Europe, North Africa, and Asia, with a "longitude-latitude" grid; the place-names here are referred to by a series of Roman numerals. The map is signed "Fantina Polo" with the year "1329." Very similar gridded configurations are found in document 8 ("Moreta Polo Map 1") and document 14 ("Moreta Polo Map 2").
References: Black, 18 and fig. 3; LAR, 5.

Document Number: 7

Title: "Fantina Polo Map 2"

Description: A map depicting East Asia, a strait, and a peninsula with a chain of islands; the drawing is set in an oval frame. A series of Roman numerals refer to a set of toponyms. It is signed "Fantina Polo" with the year "1329." Document 9 ("Lorenzo Polo Chronicle) includes a variant of the text that appears here; also note document 6 ("Fantina Polo Map 1").

References: LAR, 4.

Document Number: 8

Title: "Moreta Polo Map 1"

Description: A map covering Europe, North Africa, and Asia, with a "longitude-latitude" grid; the map is signed "Moreta Polo" with the year "1338." A very similar configuration is found in document 6 ("Fantina Polo Map 1") and document 14 ("Moreta Polo Map 2").

References: Black, 16–17 and fig. 2.

Document Number: 9

Title: "Lorenzo Polo Chronicle"

Description: A text concerning the Polo family and concerning manuscripts left by "Rugerio Sanseverino"; the text is signed "Lorenzo Polo, Protonotario, Cajatia, 1556" at the bottom of the last page, while other passages in text are signed "Carlo Sperano," "Fantina Polo," and "Moreta Polo."

References: Black, 21–25 and figs. 5 and 6; LAR, 12.

Document Number: 10

Title: "Map of the New World"

Description: On the recto, a map of Europe, North Africa, and North and South America; here the Americas are labeled "Columbia Septentrionalis" and "Columbia Meriodionalis." There is accompanying text in the form of a letter that is addressed to "Elisabetta Feltro della Rovere Sanseverino" and signed "Guido Spinola"; below the signature, one reads: "Cagliari, 20 October 1524." On the verso, there is a text mentioning Antilla and the explorer Hernando Cortez.

References: LAR, 11.

Document Number: 11

Title: "Columbus Map"

Description: A map of the New World, with a brief text; the text is signed "Giani [?] Roberto Sanseverino" and labeled "Casatia, 9 January 1620."

References: LAR, 10.

Document Number: 12
Title: "Spinola Chronicle"
Description: A text in the form of a letter to "Elisabetta Feltro della Rovere Sanseve-
  rino" and signed "Guido Spinola," along with the words "Cagliari, 25 November
  1524."
References: LAR, 8.

Document Number: 13—*missing*
Title: "Keynote to Pantect Map"
Description: A text describing a voyage by Marco Polo to a chain of islands and a
  large peninsula in the Far East; this document is reproduced and discussed by
  Bagrow in his article but is now missing from the collection.
References: Bagrow, 5 and fig. 4.

Document Number: 14—*missing*
Title: "Moreta Polo Map 2"
Description: On the recto, a map of Asia, with an oval cartouche containing an inscription in Italian; on the verso, a map covering Europe, North Africa, and Asia,
  with a "longitude-latitude" grid. The map is signed "Moretta Polo" and has an
  inscription mentioning Antilla; this document is discussed by Bagrow in his article but is now missing from the collection. Very similar gridded configurations
  are found in document 8 ("Moreta Polo Map 1") and document 6 ("Fantina Polo
  Map 1").
References: Bagrow, 4 and figs. 2 and 3.

Appendix 2:
A Partial Genealogy
of the Rossi Family

APPENDIX 2: A PARTIAL GENEALOGY OF THE ROSSI FAMILY

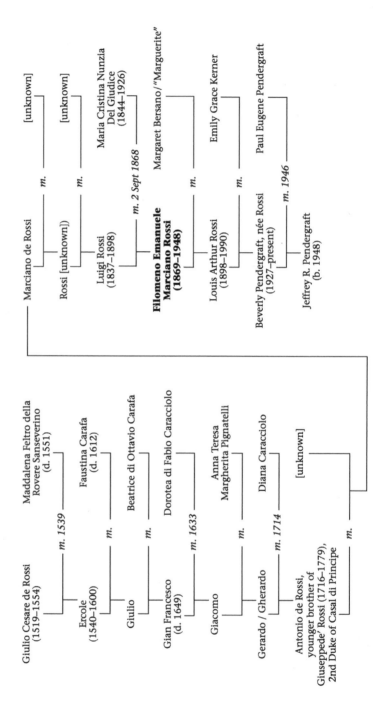

Appendix 3:
Genealogy of the Family of Marco Polo the Traveller

APPENDIX 3: A GENEALOGY OF THE POLO FAMILY
Genealogy of the Family of Marco Polo the Traveller

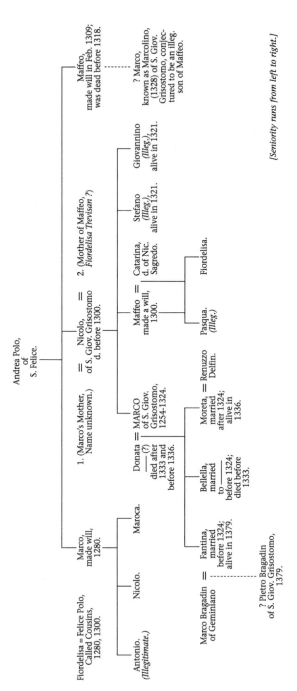

Andrea Polo,
of
S. Felice.

Fiordelisa = Felice Polo,
Called Cousins,
1280, 1300.

Antonio.
(Illegitimate.)

Nicolo.

Maroca.

Marco,
made will,
1280.

1. (Marco's Mother,
Name unknown.)

Donata = MARCO
— (?)   of S. Giov.
died after Grisostomo,
1333 and   1254-1324.
before 1336.

Marco Bragadin = Fantina,
of Geminiano   married
                before 1324;
                alive in 1379.

? Pietro Bragadin
of S. Giov. Grisostomo,
1379.

Bellella,
married
to ——
before 1324;
died before
1333.

Moreta, = Renuzzo
married   Delfin.
after 1324;
alive in
1336.

Nicolo,   =
of S. Giov. Grisostomo
d. before 1300.

2. (Mother of Maffeo,
Fiordelisa Trevisan ?)

Maffeo =
made a will,
1300.

Pasqua.
(Illeg.)

Catarina,
d. of Nic.
Sagredo.

Fiordelisa.

Stefano
(Illeg.),
alive in 1321.

Giovannino
(Illeg.),
alive in 1321.

Maffeo,
made will in Feb. 1309;
was dead before 1318.

? Marco,
known as Marcolino,
(1328) of S. Giov.
Grisostomo, conjec-
tured to be an illeg.
son of Maffeo.

[Seniority runs from left to right.]

Adapted from Henry Yule, ed. and trans., *The Book of Ser Marco Polo the Venetian concerning the Kingdoms and Marvels of the East*, 2 vols. 2nd ed., rev. (London: John Murray, 1875).

Notes

INTRODUCTION

1. Leo Bagrow, "The Maps from the Home Archives of the De-
    scendants of a Friend of Marco Polo," *Imago Mundi* 5 (1948):
    3–13.
2. Ibid., 3.
3. Ibid., 3.
4. Currently curator of maps at the Boston Public Library,
    Grim was at the time specialist in cartographic history at
    the Library of Congress. Hébert is chief of the Geography
    and Map Division of the Library of Congress.
5. Benjamin B. Olshin, "The Mystery of the 'Marco Polo'
    Maps: An Introduction to a Privately-Held Collection of
    Cartographic Materials Relating to the Polo Family," *Terrae
    Incognitae* 39 (2007): 1–23.
6. For some background on the Vinland Map and the debates
    surrounding it, see, for example, R. A. Skelton, Thomas E.
    Marston, and George D. Painter, *The Vinland Map and the
    Tartar Relation*, 2nd ed. (New Haven, CT: Yale University
    Press, 1995); Paul Saenger, "Vinland Re-read," *Imago Mundi*
    50 (1998): 199–202; and Kirsten A. Seaver, *Maps, Myths and
    Men* (Stanford, CA: Stanford University Press, 2004).
7. In addition to Jeffrey R. Pendergraft, these two other
    relatives—Victor Rossi and Richard Rossi—were extremely
    helpful in providing details and materials about their
    ancestors and extended family. Victor and Richard Rossi's
    grandfather was Florindo Lorenzo Rossi (later, in America,
    "Fred L. Rossi"), a brother of Marcian Rossi.
8. One of the most recent critiques concerning the veracity
    of the Polo narrative is by an English historian and curator
    of Chinese collections at the British Library; see Frances

Wood, *Did Marco Polo Go to China?* (London: Martin Secker & Warburg, 1995). But note the critique of Wood's work in Hans Ulrich Vogel, *Marco Polo Was in China: New Evidence from Currencies, Salts and Revenues* (Leiden: Brill, 2013); and Igor de Rachewiltz, "Marco Polo Went to China," *Zentralasiatische Studien* 27 (1997): 34–92; and idem, "Marco Polo Went to China: Additions and Corrections," *Zentralasiatische Studien* 28 (1998): 177. See, too, D. O. Morgan, "Marco Polo in China—or Not," *Journal of the Royal Asiatic Society*, 3rd ser., 6, no. 2 (July 1996): 221–225; and Peter Jackson, "Marco Polo and His 'Travels,'" *Bulletin of the School of Oriental and African Studies, University of London* 61, no. 1 (1998): 82–101. A comprehensive summarizing discussion is in John Larner, *Marco Polo and the Discovery of the World* (New Haven, CT: Yale University Press, 1999), 58–63.

9. A formal title of the Polo narrative is *Le livre des merveilles du monde* or *Le divisament du monde.*

10. Luigi Giovannini, ed., *Il Milione: Con le postille di Cristoforo Colombo* (Rome: Edizioni Paoline, 1985).

11. See Bagrow, "The Maps from the Home Archives," 3n1.

12. See the (anonymous) note in *Imago Mundi* 2 (1937): 97.

13. Martin was chief of the Library of Congress's Geography and Map Division from 1924 to 1946. The exchange of letters is described in Janice Poston, *The Rossi Collection* (Washington DC: Library of Congress, Map Division, 2 September 1992), 4.

14. Perhaps Elias Avery Lowe (1879–1969), a scholar at Oxford University, and one of the founders of the Institute for Advanced Study in Princeton.

15. On these maps, see the following: Marcel Destombes, ed., *Mappemondes A.D. 1200–1500* . . . (Amsterdam: N. Israel, 1964), 205–207; Franz R. von Wieser, *Die Weltkarte des Albertin de Virga aus dem Anfange des XV. Jahrhunderts in der Sammlung Figdor in* Wien (Innsbruck: H. Schwick, 1912); George H. T. Kimble, "The Laurentian World Map with Special Reference to Its Portrayal of Africa," *Imago Mundi* 1 (1935): 29–33; and Chet van Duzer, "Cartographic Invention: The Southern Continent on Vatican MS Urb. Lat. 274, Folios 73v–74r (c.1530)," *Imago Mundi* 59, no. 2 (June 2007): 193–222.

16. Although not a specialist in the area, Durand wrote on the history of cartography, most notably *The Vienna-Klosterneuburg Map Corpus* (Leiden: Brill, 1952).

17. In particular, Durand discusses the maps on the recto and verso of what we label here the "Moreta Polo Map 2."

18. Herbert B. Nichols, "Marco Polo's Log Comes to Light," *Christian Science Monitor*, 31 January 1936, 1–2.

19. Currently, he is Donnelley Fellow Librarian of Corpus Christi College, Cambridge.

20. William J. Wilson, *Documents 1–14: Relating to Marco Polo and Other Discoverers* (Washington DC: Library of Congress, n. d.).

21. William J. Wilson, *The Rossi Collection of Manuscript Maps and Documents* (Washington DC: Library of Congress, 1953).
22. Ibid., 8.
23. Ibid., 9ff.
24. Ibid., 12.
25. Ibid., 13.
26. Ibid., 15.
27. Ibid., 16.
28. Ibid., 46.
29. Poston, 3.
30. Lawrence Martin, "The Newly Discovered Marco Polo Map," *Annals of the Association of American Geographers* 24 (1934): 60–61.
31. Poston's *The Rossi Collection.*
32. John Black, "Marco Polo Documents Incorporated in the Felicitation Volumes of Southeast Asian Studies," in *Felicitation Volumes of Southeast-Asian Studies Presented to His Highness Prince Dhaninivat Kromamun Bidyalabh Bridhyakorn on the Occasion of His Eightieth Birthday,* ed. Prince Dhaninivat Sonakul, 2. vols (Bangkok: Siam Society, 1965), 2:13–34.

CHAPTER ONE

1. These are noted in appendix 1.
2. When I first encountered these documents, I saw that they had not been categorized at any point in a systematic way. This current categorization system, which I have devised, seems to me to allow the most clear comparative analysis of the documents and their contents.
3. Bagrow, "The Maps from the Home Archives," 5, and fig. 4.
4. Ibid., 4, and figs. 2 and 3.
5. For a careful examination of Ramusio's life, interests, and writings, see George B. Parks, "Ramusio's Literary History," *Studies in Philology* 52, no. 2 (April 1955): 127–148. We will return to Ramusio's work later in this book.
6. "The genealogy of the Polo family is neither very clear nor very sure." See p. 1 of Giovanni Orlandini, "Marco Polo e la sua familia," *Archivio veneto-tridentino* 9 (1926): 1–68. Despite the sparsity of evidence, however, Orlandini is able to provide a good summary of the Polo clan, with footnotes recounting *documenti* that mention various family members.
7. Boleslaw Szczesniak, "Marco Polo's Surname 'Milione' according to Newly Discovered Documents," *T'oung Pao,* 2nd ser., 48, no. 4/5 (1960): 447–452. See the brief summary on this question in Larner, 44.
8. Szczesniak, 450–452.
9. Ibid., 448.
10. Larner, 44.
11. "Here were the houses [*sic*] of Marco Polo, who traveled [to] the farthest regions of Asia and described them." For brief but helpful comments on

both the term *Milione* and the house(s) of Marco Polo, see pp. 302–303 of G. F. Hudson, "Marco Polo," *Geographical Journal* 120, no. 3 (September 1954): 299–311.

12. See the discussion in Henry Yule, ed. and trans., *The Book of Ser Marco Polo the Venetian Concerning the Kingdoms and Marvels of the East*, 2 vols., 2nd ed., rev. (London: John Murray, 1875), 2:507–508.

13. Note p. 223 of Luigi Villari, "A British Scholar on Marco Polo," *East and West* 5, no. 3 (October 1954): 222–226.

14. See the discussion in Italo M. Molinari, "Un articolo d'autore cinese su Marco Polo e la Cina," *Annali dell'Istituto Universitario Orientale* 42, suppl. 30 (Naples: Istituto orientale di Napoli, 1982). Also note the comments in Francis Woodman Cleaves, "A Chinese Source Bearing on Marco Polo's Departure from China and a Persian Source on His Arrival in Persia," *Harvard Journal of Asiatic Studies* 36 (1976): 181–203.

15. Larner, 41; also see Hudson, 309.

16. See the extensive treatment of the Polo family's activities in Larner, 33ff. Our discussion here of Marco Polo and his narrative follows that presented in Larner.

17. Ibid., 42–45.

18. Ibid., 46; for further details, also see the discussion on pp. 84–85 of Jackson; and Hudson, 301.

19. Larner, 46.

20. Marco Polo became a prisoner during the 1294–1299 war between the Republic of Venice and the Republic of Genoa.

21. Larner, 46–47. Jackson notes that, in fact, Polo may have had "other co-authors"; see Jackson, 84.

22. Larner, 52.

23. See Szczesniak, 447; as well as Leonardo Olschki, *Marco Polo's Asia* (Berkeley: University of California Press, 1960), 103. For a brief discussion on other mentions of Marco Polo in the fourteenth century, see Jackson, 86.

24. Jackson, 86; a more extensive discussion of the Ramusio edition is found in pp. 456–459 of J. Homer Herriott, "The 'Lost' Toledo Manuscript of Marco Polo," *Speculum* 12, no. 4 (October 1937): 456–463. For a study of Ramusio's comments on the Polo family, see Marco Polo, *The Description of the World*, trans. and annot. A. C. Moule and Paul Pelliot, 2 vols. (London: Routledge & Sons, 1938), 1:15ff. There are a many other sources on Marco Polo, including the well-known edition of Sir Henry Yule; see Yule, ed., *The Travels of Marco Polo: the Complete Yule-Cordier Edition: Including the Unabridged Third Edition (1903) of Henry Yule's Annotated Translation, As Revised by Henri Cordier, Together with Cordier's Later Volume of Notes and Addenda (1920)*, 2 vols. (New York: Dover Publications, 1993). Another work that looks at the character of Marco Polo and his narrative is John Critchley's *Marco Polo's Book* (Brookfield, VT: Variorum, 1992). On Marco Polo as an inspiration to later exploration, along with a critical examina-

tion of his narrative, see Enrico Vicentini, "Il *Milione* di Marco Polo come portolano," *Italica* 71, no. 2 (Summer 1994): 145–152.

25. Larner, 69.

26. Ibid., 70.

27. Olschki, 118.

28. The best one can do is visit the places that Marco Polo recounts and that one can indeed identify clearly; see, for example, Denis Belliveau and Francis O'Donnell, *In the Footsteps of Marco Polo* (Lanham, MD: Rowman & Littlefield, 2008); and Jack Spain, Jr., *Retracing Marco Polo: A Tale of Modern Travelers Who Locate and Follow Marco Polo's Route to China and Burma* (Richmond, VA: Gilgit Press, 2003).

29. Larner, 77.

30. Ibid., 134.

31. Ibid., 134; also see Evelyn Edson, *The World Map, 1300–1492: The Persistence of Tradition and Transformation* (Baltimore: Johns Hopkins University Press, 2007), 70–71.

32. Larner, 135.

33. Larner, 61–62, notes this as good evidence that Marco Polo himself received the tablets personally on his travels.

34. Ibid., 45.

35. For a partial translation of Marco Polo's will, see Henry Yule, ed. and trans., *The Book of Ser Marco Polo the Venetian Concerning the Kingdoms and Marvels of the East*, 2 vols., 3rd ed., rev. (London: John Murray, 1903), 1:70–73.

36. Yule provides a summary of the surviving documents and mentions of the Polos; see his "Appendix C—Calendar of Documents Relating to Marco Polo and His Family," in the edition cited in the preceding note, 2:510–521. Also see the discussion in Henry H. Hart, *Venetian Adventurer: Being an Account of the Life and Times and of the Book of Messer Marco Polo* (Stanford, CA: Stanford University Press, 1942), 249ff.

37. Yule, *The Book of Ser Marco Polo*, 3rd ed., rev. (1903), 1:79–80; for a complete list of these items, see Cesare Augusto Levi, *Le collezioni veneziane d'arte e d'antichità dal secolo XIV ai nostri giorni* (Venice: Ferd. Ongania, 1900), 35–38.

38. Yule also gives a summary of "the remaining history of Marco Polo's immediate family"—see *The Book of Ser Marco Polo*, 3rd ed., rev. (1903), 1:76–79.

39. Hart's *Venetian Adventurer* seems to give the most complete summary of the extant documents that mention Marco Polo's daughters.

CHAPTER TWO

1. This name is sometimes rendered in English as Blaise.

2. Yule, *The Book of Ser Marco Polo*, 3rd ed., rev. (1903), 2:312–313.

3. Ibid., 2:314n.2. Also see the comments in Larner, 89–90.

4. Olschki, 32.
5. See p. 256 of Gerald R. Tibbetts, "The Role of Charts in Islamic Navigation in the Indian Ocean," in *The History of Cartography*, vol. 1, *Cartography in Prehistoric, Ancient, and Medieval Europe and the Mediterranean*, ed. J. B. Harley and David Woodward (Chicago: University of Chicago Press, 1987), 256-262.
6. Yule, *The Book of Ser Marco Polo*, 3rd ed., rev. (1903), 2:424.
7. Olschki, 32-33n62.
8. See Tibbetts, 256; and the brief mention in Josef W. Meri, ed., *Medieval Islamic Civilization: An Encyclopedia*, 2 vols. (New York: Routledge, 2006), 1:556. A longer discussion of *al-qunbas* is found in Fuat Sezgin, *Mathematical Geography and Cartography in Islam and Their Continuation in the Occident*, vol. 1, *Historical Presentation*, trans. Guy Moore and Geoff Sammon (Frankfurt am Main: Institute for the History of Arabic-Islamic Science, 2005), 58 and 247.
9. I wish to thank Prof. Amir Harrak for his attempt to make sense of this Arabic text.
10. On Auxacia, see William Smith, *A Dictionary of Greek and Roman Geography*, 2 vols. (Boston: Little, Brown & Co., 1854), 1:347; and Henry Lansdell, *Chinese Central Asia: A Ride to Little Tibet*, 2 vols. (London: Sampson Low, Marston & Co., 1893), 1:357.
11. This map is discussed further in chap. 5 and chap. 7 here.
12. This translation is from Piero Falchetta, *Fra Mauro's World Map, with a Commentary and Translations of the Inscriptions*, trans. Jeremy Scott (Turnhout: Brepols, 2006), 709.
13. Ibid., 717, with slight modifications in the translation.
14. A short discussion of this is found in Ronald L. Ives, "An Early Speculation Concerning the Asiatic Origin of the American Indian," *American Antiquity* 21, no. 4 (April 1956): 420-421. An early but quite extensive treatment is in Justin Winsor, *Narrative and Critical History of America*, 8 vols. (Boston: Houghton, Mifflin and Co., 1889), 1:76ff.
15. Yule, *The Book of Ser Marco Polo*, 3rd ed., rev. (1903), 2:253-254.
16. The only Italian word similar to *Bikerne* is the term *bicherna* or *biccherna*, a word used in Siena in reference to the treasury or a treasurer. But it seems unlikely that this is the meaning here.
17. See p. 381 of Lionel Giles, "Translations from the Chinese World Map of Father Ricci," *Geographical Journal* 52, no. 6 (December 1918): 367-385.
18. Ibid., 381.
19. Olschki, 341.
20. "Place inhabited by bellicose and valiant women who fight amongst themselves"; see Falchetta, *Fra Mauro's World Map*, 677.
21. Ibid., 676.
22. See p. 222n14 in Jennifer W. Jay, "Imagining Matriarchy: 'Kingdoms of Women' in Tang China," *Journal of the American Oriental Society* 116, no. 2 (April-June 1996): 220-229.

CHAPTER THREE

1. John Hessler, personal communication, 27 July 2013.
2. Ibid.
3. John Hessler, "Warping History: Mathematical Methods in Historical Cartometry—Sketching the Unknown: A Phenomenological and Computational Study of the Rossi 'Map With Ship,'" last modified 31 December 2008, http://warpinghistory.blogspot.com/2008/12/sketching-unknown -phenomenologicalandco.html.
4. This reading is the result of examinations of the document by the author, with an additional assessment of the characters by Prof. Nathan Sivin of the University of Pennsylvania.
5. See p. 564 of Marvin W. Falk, "Images of Pre-discovery Alaska in the Work of European Cartographers," *Arctic* 37, no. 4 (December 1984): 562–573.
6. Ibid., 564.
7. Note Falk, 566; also see the discussion in Thomas Suárez, *Early Mapping of the Pacific: The Epic Story of Seafarers, Adventurers, and Cartographers Who Mapped the Earth's Greatest Ocean* (Hong Kong: Periplus Editions, 2004), 46–47. For a list and brief description of a number of early maps showing lands up in the northern extremes of the Pacific Ocean, see Justin Winsor, "The Kohl Collection of Early Maps," *Harvard University Bulletin* 4, no. 4 (January 1886): 234–241.
8. Concerning the relationship between Homann's maps and the subsequent voyages by Vitus Bering, see Orcutt W. Frost, *Bering: The Russian Discovery of America* (New Haven, CT: Yale University Press, 2003), 36–40.
9. Giles, 383.
10. See Raisa V. Makarova, *Russians on the Pacific, 1743–1799*, trans. and ed. Richard A. Pierce and Alton S. Donnelly (Kingston, ON: Limestone Press, 1975). Also note the following: Leo Bagrow, "The First Russian Maps of Siberia and Their Influence on the West-European Cartography of N.E. Asia," *Imago Mundi* 9 (1952): 83–93; and L. Breitfuss, "Early Maps of North-Eastern Asia and of the Lands around the North Pacific: Controversy between G. F. Müller and N Delisle," *Imago Mundi* 3 (1939): 87–99.
11. A full account is given in Frost, *Bering*; also see Lydia T. Black, *Russians in Alaska, 1732–1867* (Fairbanks: University of Alaska Press, 2004). Much of the exploration of this area was driven by trade—see especially pp. 393–403 of Jon D. Carlson, "The 'Otter-Man' Empires: The Pacific Fur Trade, Incorporation and the Zone of Ignorance," *Journal of World-Systems Research* 8, no. 3 (Fall 2002): 390–442.
12. Pierluigi Portinaro and Franco Knirsch, *The Cartography of North America, 1500–1800* (Edison, NJ: Chartwell Books, 1987), 286–287.
13. See p. 97 of Johannes Keunig, "Nicolaas Witsen As a Cartographer," *Imago Mundi* 11 (1954): 95–110.
14. Ibid., 100.

15. Ibid., 101.
16. Ibid., 101.
17. See Yule, *The Book of Ser Marco Polo*, 3rd ed., rev. (1903), 1:15–16 and 1:34–35.
18. Interpretatione dal Tartaro in Latinu di la tabula de
    oro de autorita de possessione supra la provincia e
    adiacens peninsule e Mari Australi e
    Orientali e confini subjecti e alieni popoli que
    Klubai Imperador de Tartaria fax a Marcus
    Polo de Venecia e da la puaesto explorator fos
    remunerato cu tampto thexoro //
    facta da Rugerius Sanseverinus
19. This may be Yanju, a city mentioned in the Polo narrative and generally believed to be modern Yangzhou.
20. I am indebted to Prof. Joseph Farrell for the emended transcription of the Latin text, and I wish to thank Prof. Farrell and Dr. Marc Cohen for their discussions with me concerning the problematic interpretation of the various terms here.
21. Cassiodorus, *Variae*, 6.23.
22. The existing Latin phrase is used in a description of the posts held by a Roman official, Rufinianus Ablabius Tatianus; see Alan Cameron, *The Last Pagans of Rome* (Oxford: Oxford University Press, 2011), 137.
23. Ettore de Ruggiero, *Dizionario epigrafico di antichità romane* (Rome: L. Pasqualucci, 1895); see 1:131 for the entry with the relevant inscription. On the terms *adlectus inter consulares, adlectus inter praetorios*, etc., note James C. Egbert, Jr., *Introduction to the Study of Latin Inscriptions* (New York: American Book Co., 1896), 167.
24. See Louis John Paetow, *The Arts Course at Medieval Universities, with Special Reference to Grammar and Rhetoric* (Champaign, IL, 1910), 71; and Paul Abelson, *The Seven Liberal Arts: A Study in Mediæval Culture* (New York: Teachers' College Columbia University, 1906), 63n4.
25. See Jackson, 86.
26. Yule, *The Book of Ser Marco Polo*, 3rd ed., rev. (1903), 1:351; also see 1:351–354n2.
27. Wood, 118.
28. Yule, *The Book of Ser Marco Polo*, 3rd ed., rev. (1903), 1:15.
29. Wood, 118.
30. Ibid., 130 and 148–149.
31. Tartaria ut ait Polo est longe ab ins fortunatae centum quindecim gradi septem horas Orientem versus. Venetiis V Julius CIƆCCXCVII //
32. Concerning Marco Polo's (apparently very limited) knowledge of Japan, see Wilcomb E. Washburn, "Japan on Early European Maps," *Pacific Historical Review* 21, no. 3 (August 1952): 221–236.
33. Insulae multae sunt principalis haec . . .

34. Now in the Vatican Library (Vat.Gr.177); the manuscript contains no maps. For more on Planudes and the manuscripts of Ptolemy, see Patrick Gautier Dalché, *La Géographie de Ptolémée en Occident (IVe–XVIe siècle)* (Turnhout: Brepols, 2009), 82–86; as well as J. Lennart Berggren and Alexander Jones, *Ptolemy's "Geography": An Annotated Translation of the Theoretical Chapters* (Princeton, NJ: Princeton University Press, 2000), 49–50.
35. Larner, 140.
36. Ibid., 140.
37. Pliny the Elder, *Natural History*, 6.37.203.
38. The original of this map was in Munich but disappeared after the war. For a brief discussion of this map and its origins, see pp. 95–96 of J. H. F. Sollewijn Gelpke, "Afonso de Albuquerque's Pre-Portuguese 'Javanese' Map, Partially Reconstructed from Francisco Rodrigues' Book," *Bijdragen tot de Taal-, Land- en Volkenkunde* 151, no. 1 (1995): 76–99. Concerning Pedro Reinel, also see p. 519 of Armando Cortesão, "A Hitherto Unrecognized Map by Pedro Reinel in the British Museum," *Geographical Journal* 87, no. 6 (June 1936): 518–524; and Jean Denucé, *Les origines de la cartographie portugaise et les cartes des Reinel* (Gand: E. van Goethem, 1908), 31–36.
39. In the Lopo Homem planisphere of 1519, we find "CHIS" written on China itself; in the Miller Atlas, there is "CHIIS."
40. "L'emplacement de cette île correspond (si l'on veut) au Japon et non à la Chine." See Albert Kammerer, *La découverte de la Chine par les Portugais au XVIème siècle et la cartographie des portulans* (Leiden: E. J. Brill, 1944), 191; a similar conjecture is Denucé, 126.
41. "La carte des Reinel se termine du côté de l'est par une longue ligne tracée au simple trait, qui doit représenter, tout à fait au hasard, la côte occidentale du Nouveau-Continent." See pp. 162–163 of E. T. Hamy, "L'oeuvre géographique des Reinel et la découverte des Moluques," in *Etudes historiques et géographiques*, ed. E. T. Hamy (Paris: Ernest Leroux, 1896), 145–177.
42. Henry Harrisse, *The Discovery of North America: A Critical, Documentary, and Historical Investigation* (Amsterdam: N. Israel, 1961), 471.
43. See Alberto Magnaghi, "La prima rappresentazione delle Filippine e delle Molucche dopo il ritorno della spedizione di Magellano, nella carta construita nel 1522 da Nuño Garcia de Toreño, conservata, nella Biblioteca di S. M. Il Re in Torino," *Atti del X Congresso Geografico Italiano* 1, anno 6 (1927): 293–307.
44. See Magnaghi; and note Albert Kammerer, *La Mer Rouge, l'Abyssinie, et l'Arabie aux XVIe et XVIIe siècles et la cartographie des portulans du monde oriental*, 3 vols. (Cairo: Société Royale de Géographie d'Égypte, 1929–1952), 1:445–447, and pl. 164.
45. "On ne connaît aucune carte ou mappemonde originale englobant l'Extrême Orient entre 1492 et 1502. A cette dernière date fut dessinée la

célèbre mappemonde anonyme portugaise dite de Cantino." Kammerer, *La découverte de la Chine*, 189.

46. See Armando Cortesão and A. Teixeira da Mota, *Portugaliae Monumenta Cartographica*, 6 vols. (Lisbon: Imprensa Nacional, 1960), 1:7–13; also note Gago Coutinho, "O mais antigo mapa da América," separata do *Boletim da Sociedade de Geografia de Lisboa* 73, no. 1–3 (January–March 1955): 2–14.

47. *Portugaliae Monumenta Cartografica*, 1:9.

48. Ibid., 1:33–34 and pl. 33; also see Thomas Suárez, *Early Mapping of Southeast Asia* (Hong Kong: Periplus, 1999), 39.

49. Kammerer, *La découverte de la Chine*, 194; Fra Mauro's 1460 map, for example, has Zimpagu.

50. Ibid., 41–47; also note Armando Cortesão, "The First Account of the Far East in the Sixteenth Century: The Name 'Japan' in 1513," in *Comptes rendus du Congrès international de géographie, Amsterdam, 1938*, 2 vols. (Leiden: E. J. Brill, 1938): 2:146–152.

51. The interpretation depends on whether we read the Portuguese *tem* (the verb "to have") as *tem* (third person singular) or as *têm* (third person plural).

52. Kammerer, *La découverte de la Chine*, 198.

53. For a brief discussion of early European explorations of the Pacific, see Charles E. Nowell, "The Discovery of the Pacific: A Suggested Change of Approach," *Pacific Historical Review* 16, no. 1 (February 1947): 1–10.

54. Kammerer, *La Mer Rouge*, 1:424: "A l'est et au nord de Singapour se creuse encore un Sinus Magnus informe et considérable, avec cette différence qu'il remonte tellement au nord qu'on n'en voit pas le fond. Il constitue ainsi une mer de Chine (Chis). Mais son bord oriental redescend vers le sud, puis vers le sud-est et ne saurait pas plus qu' à la planche CII [i.e., his "1516 map of Reinel"], être assimilé à la côte américaine, que le cartographe ignore totalement. Il s'en faut de la largeur du Pacifique." Note again the similar depiction in the 1522 work of Nuño Garcia de Toreno, in Kammerer, *La Mer Rouge*, 1:445–447 and pl. 164.

55. Armando Cortesão, *Cartografia e cartógrafos portugueses dos séculos XV e XVI*, 2 vols. (Lisbon: Seara Nova, 1935), 1:270–272.

CHAPTER FOUR

1. I wish to thank Prof. Vladimiro Valerio of Università IUAV di Venezia for his help in transcribing this text from the original manuscript.

2. Prof. Valerio assisted in the transcription of this text from the original manuscript.

3. Bagrow, "The Maps from the Home Archives," figs. 2 and 3.

4. Ibid., 4.

5. This reading is the result of examinations of the document by author, with an additional assessment of the characters by Prof. Nathan Sivin.

NOTES TO PAGES 62–64

A brief translation of these individual characters—which does not give us much indication of the meaning of the overall inscription here—is as follows: 王 (*wáng* "king"), 有 (*yǒu* "to have" or "there is/are"), 百 (*bǎi* "hundred"), 人 (*rén* "person, people") or 入 (*rù* "enter, go into"), 之 (*zhī* possessive particle), and 即 (*jí* "namely, that is," etc.).

6. Armando Cortesão, *The Nautical Chart of 1424 and the Early Discovery and Cartographical Representation of America: A Study on the History of Early Navigation and Cartography* (Coimbra: University of Coimbra, 1954).

7. See pp. 3–4 of J. B. Harley, "Text and Contexts in the Interpretation of Early Maps," in David Buisseret, ed., *From Sea Charts to Satellite Images: Interpreting North American History through Maps* (Chicago: University of Chicago Press, 1990): 3–15.

8. Pliny the Elder, *Natural History*, 6.37.203.

9. Armando Cortesão's key works on this subject are "The North Atlantic Nautical Chart of 1424," *Imago Mundi* 10 (1953): 1–13; and *The Nautical Chart of 1424*. Also see his article entitled "Pizzigano's Chart of 1424," *Revista da Universidade de Coimbra* 24 (1970): 477–491, republished in *Agrupamento de Cartografia Antiga, Junta de Investigaões do Ultramar*, série separatas, no. 40 (Coimbra: University of Coimbra, 1970). There are also numerous other works concerning the question of Antilia and its identification; see, for example, Roberto Almagià, "Intorno all'Antilia, isola legendaria dell'Atlantico," *Rivista Geografica Italiana* 53 (1936): 249–250; William H. Babcock, "Antillia and the Antilles," *Geographical Review* 9 (1920): 109–124; G. R. Crone, "The Origin of the Name Antillia," *Geographical Journal* 91 (1938): 260–262; and Giandomenico Serra, "Da Altino alle Antille, appunti sulla fortuna e sul mito del nome 'Altilia,' 'Attilia,' 'Antilia' . . . ," *Studii Italiene* (Bucharest) 2 (1935): 25–99.

10. See Cortesão, *The Nautical Chart of 1424*, 69.

11. António Galvão, *Tratado dos descobrimentos* (Lisbon: Imp. João Barreira, 1563); reprint, 3rd ed. (Porto: Livraria Civilização, 1944).

12. Pedro de Medina, *Libro de grandezas y cosas memorables de España* (Sevilla: Imp. Domenico de Robertis, 1548).

13. Ibid., chap. 42; this translation is from Cortesão, *The Nautical Chart of 1424*, 71, with some slight modifications.

14. Cortesão, *The Nautical Chart of 1424*, 71. The *Monarquia Lusitana*, written right at the end of the sixteenth century and the beginning of the seventeenth century, repeats Medina's account exactly. But in a marginal note, it refers to a Ptolemy map of 1523. See Frei Bernardo de Brito, *Monarquia Lusitana*, 8 pts. (Lisbon: Imprensa Nacional–Casa da Moeda, 1973–1988), 2nd pt., 282r.

15. On Zacuto, see Luís de Albuquerque, "Abraham Zacuto et l'Almanach Perpetuum," *Estudos de História* 5 (1977): 51–80; António Barbosa, *O almanach perpetuum de Abraham Zacuto e as tábuas náuticas portuguesas* (Coimbra: Imp. da Universidade, 1928); Francisco Cantera Burgos, *El Judío Salmantino*

Abraham Zacut (Madrid: Academia de Ciencias, 1931); Berthold Cohn, *Der Almanach Perpetuum des Abraham Zacuto* (Strassburg: K. J. Trübner, 1918); and Abel Fontoura da Costa, *L'Almanach Perpetuum de Abraham Zacut (Leiria, 1496)* (Lisbon: Seara Nova, 1935).

16. The work itself is available in the following: Joseph Vizinho, trans., *Almanach Perpetuum Celestium Motuum (Radix 1473)* . . . , a facsimile of the 1496 Leiria edition in the Biblioteca de Evora; this is vol. 6 of Joaquim Bensaúde, ed., *Histoire de la science nautique portugaise à l'époque des grandes découvertes*, 7 vols. (Munich: Carl Kuhn, 1914–1917). A facsimile of the 1496 edition in the Augsbourg library is found in vol. 3 of this collection. A facsimile of the copy in the Biblioteca Nacional in Lisbon is the *Almanach Perpetuum de Abraão Zacuto*, with an introduction by Luís de Albuquerque (Lisbon: Imprensa Nacional–Casa da Moeda, 1986). Note that Vizinho (José Vizinho, Jewish astrologer and doctor) leaves out chap. 9 of the Canons, the chapter with the passage in question; thus, of the modern facsimiles and editions only Cantera Burgos's *Abraham Zacut: Siglo XV* (Madrid: M. Aguilar, 1935) includes it, as he is working from the Hebrew manuscript in Lyon. In addition, the Spanish version of Juan Selaya included this important chapter—his work was also published by Cantera Burgos, in *El Judío Salmantino Abraham Zacut* (Madrid: Academia de Ciencias, 1931); see pp. 133 and 232. The Selaya version actually has "17 grados e 20 minutos" rather than the 17½ degrees we find here.

17. "Ten entedido que la longitud de las ciudades desde se cuenta de occidente para oriente, y, como nosotros estamos más próximos al occidente, hemos comezado a contar por el lado occidental. Otros calcularon la longitud de las ciudades desde una isla que hay en occidente, del qual ella dista 17 grados y medio, por lo que incurrieron en error la mayor parte delas tabla." This Spanish translation of the Hebrew manuscript in Lyon is from Burgos, 142.

18. See p. 180 of Zhang Zhishan, "Columbus and China," *Monumenta Serica* 41 (1993): 177–187. The original passage begins: "E de la isla de Antilla, que vosotros llamáis de Siete Ciudades, de cual tenemos noticia, hasta la nobilísima isla de Cipango, hay 10 espacios, que son 2,500 millas." The Toscanelli letter is found in Bartolomé de las Casas, *Historia delas Indias*, lib. 1, cap. 12; see Bartolomé de las Casas, *Historia de las Indias*, 3 vols., ed. Agustín Millares Carlo (Mexico City: Fondo de Cultura Económica, 1951), 1:62–65. The letter is also found in the writings of Ferdinand Columbus; see Benjamin Keen, ed. and trans., *The Life of the Admiral Christopher Columbus by His Son Ferdinand* (New Brunswick, NJ: Rutgers University Press, 1959), 19–22. Also note William H. Babcock, *Legendary Islands of the Atlantic: A Study in Medieval Geography* (New York: American Geographical Society, 1922), 69; and see "Carta de Toscanelli a Fernão Martins, de 25 de Junho de 1474," in Carlos Malheiro Dias, ed., *História da Colonização Portuguesa do Brasil*, 3 vols. (Oporto: Litografia Nacional, 1921–1924), 1:1xxx.

19. See W. G. L. Randles, "Le projet asiatique de Christophe Colomb devant la science cosmographique portugaise et espagnole de son temps," separata da revista *Islenha* (Funchal, Madeira) 5 (July–December 1989): 73–88.

20. E. G. Ravenstein, *Martin Behaim: His Life and His Globe* (London: George Philip & Son, 1908), 77; also see Babcock, *Legendary Islands of the Atlantic*, 71. There are a number of references to the tale of the "Island of the Seven Cities"; see, for example, William H. Babcock, "The Island of the Seven Cities," *Geographical Review* 7, no. 2 (February 1919): 98–106; also note pp. 32–37 of Jaime Cortesão, "The Pre-Columbian Discovery of America," *Geographical Journal* 89, no. 1 (January 1937): 29–42; as well as pp. 457–458 of G. R. Crone, "The Alleged Pre-Columbian Discovery of America," *Geographical Journal* 89, no. 5 (May 1937): 455–460.

21. This text is very difficult to decipher, since the only surviving image is in Bagrow's article and is not very clear; see Bagrow, "The Maps from the Home Archives," 4.

22. Ibid., 11.

CHAPTER FIVE

1. See Constancio Gutiérrez, *Trento, un problema: La última convocación del Concilio (1552–1562)* (Madrid: Universidad Pontificia Comillas, 1995), 481n1.

2. Dr. Carla Weinberg assisted in these translations.

3. Cajatia (or Caiatia) is the Latinate version of Caiazzo, a city in the region of Campania in Italy.

4. Vittorio Spreti, *Enciclopedia storico-nobiliare italiana . . .* , 6 vols. (Milan: Ed. Enciclopedia storico-nobiliare italiana, 1928–1932), 6:104.

5. Giovanni Battista Ramusio, *Navigationi et Viaggi*, 3 vols. (Venice: Giunti, 1563–1606, facs. ed., Amsterdam: Theatrum Orbis Terrarum, 1967–1970).

6. See Wood, 111–120.

7. As noted in chap. 1, for a discussion of the Ramusio version, see Herriott, 456–459; as well as Parks, "Ramusio's Literary History." Concerning the *Navigationi et Viaggi*, note George B. Parks, "The Contents and Sources of Ramusio's *Navigazioni*," *Bulletin of the New York Public Library* 59, no. 6 (June 1955): 279–313; and the source that Parks uses (and expands upon): Antonio del Piero, "Della vita e degli studi di Gio. Battista Ramusio," *Nuovo Archivo Veneto*, n.s., 4, no. 1 (1902): 5–112.

8. Ramusio, vol. 2, fol. 51r. This English translation is from Henry R. Wagner, *Spanish Voyages to the Northwest Coast of America* (San Francisco: California Historical Society, 1929), 126; also note 355–361. See the brief comments on this passage in Yule, *The Book of Ser Marco Polo*, 3rd ed., rev. (1903), 2:266.

9. Concerning the map of Fra Mauro, see Falchetta, *Fra Mauro's World Map*; and Angelo Cattaneo, *Fra Mauro's Mappa Mundi and Fifteenth-Century*

Venice (Turnhout: Brepols, 2011). Also see Giacinto Placido Zurla, *Il mappamondo di Fra Mauro Camaldolese* (Venice, 1806); and note Tullia Gasparrini Leporace, *Il mappamondo di Fra Mauro* (Rome: Instituto Poligrafico dello Stato, Libreria dello Stato, 1954). Although dated in some respects, Zurla's work gives a thorough description of the map and an extensive discussion of its milieu, and Leporace's study provides an excellent color facsimile of the map and transcriptions of the legends.

10. Falchetta, *Fra Mauro's World Map*, 457; also note Zurla, *Il mappamondo di Fra Mauro*, 38.

11. See Henry R. Wagner, *The Cartography of the Northwest Coast of America to the Year 1800*, 2 vols. (Berkeley: University of California Press, 1937), 1:355–361; as well as Woodward, "The Forlani Map of North America," *Imago Mundi* 46 (1994): 29–40.

12. See p. 404 of Christian Sandler, "Die Anian-Strasse und Marco Polo," *Zietschrift der Gesellschaft für Erdkunde* 29 (1894): 401–408; also note A. E. Nordenskiöld, "The Influence of the 'Travels of Marco Polo' on Jacobo Gastaldi's Maps of Asia," *Geographical Journal* 13, no. 4 (April 1899): 396–406.

13. Sandler, 404–405 and 404n3.

14. See pp. 31–33 and fig. 3 of Errol Wayne Stevens, "The Asian-American Connection: The Rise and Fall of a Cartographic Idea," *Terrae Incognitae* 21 (1989): 27–39.

15. Sandler, 405.

16. Sandler, 406–408; Zurla, *Il mappamondo di Fra Mauro*, 114, states that Fra Mauro's map includes Kamchatka: "[L']ultimo confine della Siberia al Nord-Est sia espresso nella vicina curva piegatura del Continente settentrionale, e il golfo che la separa dalle regioni prominenti di Mongul alla Chinese Tartaria spettanti, debbasi riputare il mare di Kamtschatka, o golfo dell'Amur." ["The last border of Siberia to the northeast may be represented by the adjoining bent curve of the northern continent, and the gulf which separates it from the prominent regions of Mongul, belonging to Chinese Tartary, so that one ought to consider [this area] the sea of Kamchatka, or the Gulf of Amur."]

17. Johannes Schöner, *Opusculum Geographicum ex diversorum libris ac cartis* . . . (Nuremberg: Johann Petreius, 1533), pt. 1, chap. 1.

18. Schöner, pt. 2, chap. 20. This passage is cited and discussed in Alexander von Humboldt, *Examen critique de l'histoire de la géographie du nouveau continent, et des progrès de l'astronomie nautique aux quinzième et seizième siècles*, 5 vols. (Paris: Gide, 1836–1839), 5:167–175; as well as Franz R. von Wieser, *Magalhâes-Strasse und Austral-Continent auf den Globen des Johannes Schöner* (Innsbruck: Verlag der Wagner'schen Universitäts-Buchhandlung, 1881): 81–82.

19. This passage is from bk. 1, chap. 1, p. 8, of Frei Gregorio García, *Origen de los Indios de el Nuevo Mundo e Indias Occidentales* (Madrid: F. Martinez

Abad, 1729). Note that this is the 12th edition of the original Valencia 1607 publication and includes various additions to the text (here indicated by braces). For more on this work, see Alain Desreumaux and Francis Schmidt, eds., *Moïse Géographie: Recherches sur les représentations juives et chrétiennes de l'espace* (Paris: J. Vrin, 1988), 164.

20. García, 8.
21. A transcribed list of place-names from this Oronce Finé map can be found in Charles-Victor Langlois, "Étude sur deux cartes d'Oronce Fine de 1531 et 1536," *Journal de la Société des Américanistes*, n.s., 14–15 (1922): 83–97.
22. For a discussion of this vital historical question, see Felipe Fernández-Armesto, *Columbus* (Oxford: Oxford University Press, 1991), 27ff. and 95.
23. Ibid., 95.
24. Ibid., 32.
25. Ibid., 30.
26. Ibid, 30. Also see George E. Nunn, *The Columbus and Magellan Concepts of South American Geography* (Glenside, PA, 1932), 15–23; Nunn provides various pieces of evidence for Columbus's transatlantic, Asiatic beliefs and intentions.
27. Peter Martyr d'Anghera, *De Orbe Novo*, 2 vols., ed. and trans. Francis Augustus McNutt (New York: G. P. Putnam's Sons, 1912), 1:65. For the Ruysch map, see Nunn, *The Columbus and Magellan Concepts*, 30–31. Concerning Martyr's idea that the coasts of South America were part of Asia, see Suárez, *Early Mapping of Southeast Asia*, 97.
28. See the discussion in Duane W. Roller, *Through the Pillars of Herakles: Greco-Roman Exploration of the Atlantic* (New York: Routledge, 2006), 27; Babcock also notes a number of early mentions by Avienus and other of seaweed and other marine obstructions; see Babcock, *Legendary Islands of the Atlantic*, 27–32.

CHAPTER SIX

1. The map is entitled *Taboas geraes de toda a navegação, divididas e emendadas por Dom Ieronimo de Attayde com todos os portos principaes das conquistas de Portugal delineadas por Ioão Teixeira cosmographo de Sua Magestade, anno de 1630* and was drawn by João Teixeira Albernaz (fl. 1602–1648).
2. See Washburn, 236.
3. For an interpretation of Antilla as a representation of Japan, see Robert H. Fuson, *Legendary Islands of the Ocean Sea* (Sarasota: Pineapple Press, 1995), 202.
4. The 1494 Treaty of Tordesillas set up a division of recently discovered lands across the Atlantic Ocean. A meridian line some 370 leagues west of the Cape Verde Islands served as the demarcation between Portuguese and Spanish claims: those to the east of this line were granted to the Portuguese, and those to the west were given to Spain. See Jesús Varela Marcos,

El Tratado de Tordesillas en la política atlántica castellana (Valladolid: Universidad de Valladolid, 1997); and Varela Marcos, ed., *El Tratado de Tordesillas en la cartografía histórica* (Valladolid: Junta de Castilla y León–V Centenario Tratado de Tordesillas, 1994); also note Paul Gottschalk, *The Earliest Diplomatic Documents on America: The Papal Bulls of 1493 and the Treaty of Tordesillas Reproduced and Translated with Historical Introduction and Explanatory Notes* (Berlin: P. Gottschalk, 1927), and Samuel Edward Dawson, *The Lines of Demarcation of Pope Alexander VI and Treaty of Tordesillas, A.D. 1493 and 1494* (Ottawa: James Hope, 1899).

5. Diodorus Siculus, *Historical Library*, 5.19–20.

6. Plutarch, *Life of Sertorius*, 8.2.

7. Pseudo-Aristotle, *On Marvellous Things Heard*, 84.

8. See Keen, 25; also note p. 79 of James Romm, "New World and '*novos orbes*': Seneca in the Renaissance Debate over Ancient Knowledge of the Americas," in *The Classical Tradition and the Americas*, vol. 1, *European Images of the Americas and the Classical Tradition*, pt. 1, ed. Wolfgang Haase and Meyer Reinhold (Berlin: Walter de Gruyter, 1994), 77–116.

9. As in the case of the "Map with Ship," the vellum of the "Columbus Map" seems to bear some underwriting, as in a palimpsest, but the writing is not readily decipherable.

10. The year of 1535 did see, however, explorations of what is today the province of Quebec by the explorer Jacques Cartier.

CHAPTER SEVEN

1. It should be noted, however, that various new editions of Ptolemy's coordinates included additional maps—the *tabulae novae*—that often covered parts of the world that were not known or had been poorly understood in Ptolemy's time, e.g., Scandinavia, the Baltic regions, and so on.

2. Grid systems apparently had been used in the ancient world by the Chinese in their maps, but not by any Western culture. It should be noted, however, that the Chinese grids did not represent longitude and latitude; see Norman J. W. Thrower, *Maps and Civilization: Cartography in Culture and Society*, 3rd ed. (Chicago: University of Chicago Press, 2008), 30.

3. For a clear and concise examination of the history of the transmission of Ptolemy's *Geography*, see Berggren and Jones, 42ff.

4. See Jay A. Levenson, ed., *Circa 1492: Art in the Age of Exploration* (New Haven, CT: Yale University Press; Washington DC: National Gallery of Art, 1991), 228–229; also note Aubrey Diller, "The Greek Codices of Palla Strozzi and Guarino Veronese," *Journal of the Warburg and Courtauld Institutes* 24, no. 3/4 (July–December 1961): 313–321, and see Berggren and Jones, 52.

5. The figure of 1,378 islands also appears in a third-century AD Greek work; see p. 205 of Robert H. Hewsen, "The *Geography* of Pappus of Alexandria: A Translation of the Armenian Fragments," *Isis* 62, no. 2 (Summer 1971): 186–207.

6. Other versions of the Polo narrative have "12,700" islands in the Indian Ocean—see Henry Vignaud, *Toscanelli and Columbus* (London: Sands & Co., 1902), 211–213 and n. 208.

7. On Pipino's text, see Consuelo Wager Dutschke, *Francesco Pipino and the Manuscripts of Marco Polo's Travels* (Ann Arbor, MI: UMI Dissertation Services, 1993).

8. Studies—both serious and spurious—of this legend abound; more serious works include Richard Hennig, "Der Buddhistenmönch Hui-Schen in 'Fusang,'" in *Terrae Incognitae: Eine Zusammenstellung und kritische Bewertung der wichtigsten vorcolumbischen Entdeckungsreisen an Hand der darüber vorliegenden Originalberichte*, 4 vols., ed. Richard Hennig (Leiden: E. J. Brill, 1936–1939), 1:33–41; and the sober analysis in Eugene R. Fingerhut, *Who First Discovered America? A Critique of Pre-Columbian Voyages* (Claremont, CA: Regina Books, 1984). See, too, the discussion in Edward H. Schafer, "Fusang and Beyond: The Haunted Seas to Japan," *Journal of the American Oriental Society* 109, no. 3 (July–September, 1989): 379–399; and in Charles Holcombe, "Trade-Buddhism: Maritime Trade, Immigration, and the Buddhist Landfall in Early Japan," *Journal of the American Oriental Society* 119, no. 2 (April–June, 1999): 280–292.

9. See the brief dimissal on p. 404 of L. Carrington Goodrich, "China's First Knowledge of the Americas," *Geographical Review* 28, no. 3 (July 1938): 400–411.

10. The story of Hui Shen's travels appears in the early seventh-century AD work *Liang shu* (梁書, "History of the Liang Dynasty") by Yao Silian (姚思廉), and in the later seventh-century *Nan shi* (南史, "History of the Southern Dynasties"), compiled by Li Yanshou (李延壽). The *Nan shi* was copied and amended by Ma Duanlin (馬端臨), in his early fourteenth-century *Wen xian tong kao* (文獻通考, "Comprehensive Studies in Literature"). This work, in a section entitled "Si yi kao" (四裔考, "Investigation of the Four Frontiers"), recounts Hui Shen's story. The story also appears in a slightly different form in the *Liang si gong ji* (梁四公紀, "Memoir of the Four Gentleman of Liang"), a work dating from the late seventh century and ascribed to Zhang Yue (張說). Fusang is also discussed in the famous *Shan hai jing* (山海經, "Classic of Mountains and Seas"), an early Chinese work that describes various wonders of geography and ethnography.

11. The traditional Chinese unit of distance *li* (里) has had various equivalents throughout history; the current conversion is 1 *li* = 500 meters.

12. See Jay, 22 and esp. n. 14; also note Bagrow, "The Maps from the Home Archives," 13.

13. Johann Georg Kohl, *Asia and America: An Historical Disquisition Concerning the Ideas Which the Former Geographers Had about the Geographical Relation and Connection of the Old and New World* (Worcester, MA: American Antiquarian Society, 1911), 44–45.

14. Joseph de Guignes, "Recherches sur les navigations des Chinois du côté de l'Amérique, & sur quelques peuples situés à l'extrémité orientale de l'Asie," *Mémoires de Littérature, Tirés des Registres de l'Académie Royale des Inscriptions et Belles-Lettres* 28 (1761): 503–525.

15. A similar claim that Asians arrived very early in the New World is found in an oddly titled nineteenth-century book: John Ranking, *Historical Researches on the Conquest of Peru, Mexico, Bogota, Natchez, and Talomeco, in the Thirteenth Century, by the Mongols, Accompanied with Elephants, and the Local Agreement of History and Tradition, with the Remains of Elephants and Mastodontes, Found in the New World* . . . (London: Longman, Rees, Orme, Brown & Green, 1827). A summary review is found in *Literary Gazette and Journal of Belles Lettres, Arts, Sciences, etc.*, no. 760 (Saturday, 13 August 1831), 513–515.

16. Heinrich Julius Klaproth, "Recherches sur le pays de Fou Sang mentionné dans les livres Chinois et pris mal àpropos pour une partie de l'Amérique," *Nouvelles annales des voyages, et des sciences géographiques* . . . , 2nd ser., 21 (1831): 53–68.

17. Charles Hippolyte de Paravey, *L'Amérique sous le nom de pays de Fou-sang* (Paris: Treuttel et Wurtz, 1844); and de Paravey, *Nouvelles preuves que le pays du Fou-sang mentionné dans les livres chinois est l'Amérique* (Paris: Édouard Bautruche, 1847).

18. Hubert Howe Bancroft, *The Native Races of the Pacific States of North America*, 5 vols. (New York: D. Appleton & Co., 1874–1876).

19. Samuel Wells Williams, "Notices of Fu-Sang, and Other Countries Lying East of China, Given in the Antiquary Researches of Ma Twan-Lin," *American Oriental Society Journal* 11 (1882–1885): 89–116.

20. Charles Godfrey Leland, *Fusang; or, The Discovery of America by Chinese Buddhist Priests in the Fifth Century* (New York: J. W. Bouton, 1875).

21. Bagrow, "The Maps from the Home Archives," 6.

22. "The Anthropology Section had assembled some first-rate documents regarding colonial history and that of Louisiana in particular. Next to them, [there were] some very disputable pieces. Not much can be said about what the map of Taddeo Visco of Genoa might depict; with good reason, one doubts its authenticity, and the letter that accompanies it seems equally suspect." *Journal de la Société des Américanistes*, n.s., 3, no. 1 (1906): 85.

23. The identity of this person is not clear; he may be the same as the naturalist from this period Sylvain Eichard, who was a member of the Société de Géographie in Paris. See *La Géographie: Bulletin de la Société de géographie* 2 (1900): 16; as well as 8 (1903): 415–417.

24. This was Baron Étienne Hulot (1857–1918), who served as secretary general from 1897 to 1918.
25. *Italica: The Quarterly Bulletin of the American Association of Teachers of Italian* 6, no. 4 (December 1929): 125.
26. Wilson, *Documents 1–14*; and Wilson, *The Rossi Collection*.
27. Juan G. Frivaldo was a governor of a province in the Philippines.
28. Bagrow, "The Maps from the Home Archives," 12.
29. Wood, 111–112.
30. See Bagrow, "The Maps from the Home Archives," 11–12.
31. See Seaver. The debates concerning the Vinland Map continue unabated; see, for example, P. D. A. Harvey, "The Vinland Map, R. A. Skelton and Josef Fischer," *Imago Mundi* 58, no. 1 (February, 2006): 95–99.
32. See pp. 170–171 of Luis Madureira, "The Accident of America: Marginal Notes on the European Conquest of the World," *New Centennial Review* 2, no. 1 (Spring 2002): 117–181.
33. Concerning the "Straits of Anian," see, for example: Robert R. Owens, "The Myth of Anian," *Journal of the History of Ideas* 36, no. 1 (January 1975): 135–138; Godfrey Sykes, "The Mythical Straits of Anian," *Bulletin of the American Geographical Society* 47, no. 3 (1915): 161–172; George E. Nunn, *Origin of the Strait of Anian Concept* (Philadelphia: Priv. print., 1929); and Sandler.
34. An extensive discussion is found in Pedro de Novo y Colson, *Sobre los viajes apócrifos de Juan de Fuca y de Lorenzo Ferrer Maldonado* (Madrid: Fortanet, 1881), 49ff.; also see the more recent W. Michael Mathes, *Fakes, Frauds, and Fabricators: Ferrer Maldonado, De Fuca, and De Fonte: The Strait of Anian, 1542–1792* (Fairfield, WA: Ye Galleon Press, 1999).
35. Related to this is the early concept of a connection between Asia and northwestern North America. An example of a work on this subject is José Torrubia, *I moscoviti nella California, o sia, dimostrazione della verità del passo all'America settentrionale nuovamente scoperto dai russi, e di quello anticamente praticato dalli popolatori, che vi trasmigrarono dall'Asia. Dissertazione storico-geografica del padre F. Giuseppe Torrubia* . . . (Rome: Generoso Salomoni, 1759). In this book, Torrubia argued that the Native Americans had migrated to North America from Asia, a theme directly connected to the discussions in some of the Rossi documents.
36. Despite the use of such toponyms, using the Polo narrative to actually construct a map is difficult for several reasons; most notably, the narrative includes few details of distances and directions, and the way in which Asian place-names are rendered sometimes makes it unclear what locale is being discussed.
37. Larner, 147.
38. Ibid., 148; also see 191ff., where Larner lists in chronological order the world maps of the fifteenth century that include material from the Polo narrative.

39. Ibid., 149; for more on Behaim, see Ravenstein.
40. Concerning the influence of the Polo narrative on Columbus, see, for example, Nicolás Wey-Gómez, *The Tropics of Empire: Why Columbus Sailed South to the Indies* (Cambridge, MA: M.I.T. Press, 2008), 137–141, 409, and 417–419.
41. Ibid., 402.
42. Yule, *The Book of Ser Marco Polo*, 3rd ed., rev. (1903), 1:110n. Also see the brief discussion of maps in relation to Marco Polo's travels in ibid., 1:135–137; and note the early but interesting treatment of maps and Marco Polo on p. 351 of R. H. Major, "Native Australian Traditions," *Transactions of the Ethnological Society of London* 1 (1861): 349–353.
43. Yule, *The Book of Ser Marco Polo*, 3rd ed., rev. (1903), 1:110n.
44. Ibid.
45. This translation is from Falchetta, *Fra Mauro's World Map*, 62, with some slight alterations; I wish to thank Prof. Carla Weinberg for her assistance with the original Italian text, which can be found in Ramusio, 2:17.
46. See the discussion in Olschki, 33n62; and the comments in Falchetta, *Fra Mauro's World Map*, 62.
47. Cattaneo, 67.
48. Ibid., 68–69.
49. For an example of this idea in the context of maps and exploration, see Romm, 79. An interesting earlier discussion concerning the connection between classical works and the discovery of the Americas is found in Peter de Roo, *History of America before Columbus according to Documents and Approved Authors*, vol. 1, *American Aborigines* (Philadelphia: J. B. Lippincott, 1900), 117ff.
50. Ramusio, 2:17, author's translation.
51. Note the discussion in Roller, 22ff. Also note Cattaneo, 117–118 and 272–273.
52. See Piero Falchetta, "Giambattista Ramusio, le mappe cinesi di Marco Polo e il mappamondo di Fra Mauro," in *Cartografi veneti: Mappe, uomini e istituzioni per l'immagine e il governo del territorio*, ed. Vladimiro Valerio (Padova: Editoriale Programma, 2007), 115–118. A briefer version of this argument is given in Falchetta, *Fra Mauro's World Map*, 62–63.
53. Giacinto Placido Zurla, *Di Marco Polo e degli altri viaggiatori veneziani più illustri dissertazioni del P. Ab. D. Placido Zurla . . .*, 2 vols. (Venice: Gio. Giacomo Fuchs co' tipi Picottiani, 1818).
54. Giacinto Placido Zurla, *Sulle antiche mappe idro-geografiche lavorate in Venezia* (Venice: Tipografia Picotti, 1818).
55. See Zurla, *Sulle antiche mappe*, 84–85.
56. Ibid., 88.
57. Bagrow, "The Celebrations of the 700th Anniversary of Marco Polo's Birth at Venice," *Imago Mundi* 12 (1955): 139–140.

58. One particularly notable study on the use of radiocarbon dating of manuscript materials is Rainer Berger et al., "Radiocarbon Dating of Parchment," *Nature* 235 (21 January 1972): 160–161.

59. See the extensive technical discussion in Richard L. Brunelle and Kenneth R. Crawford, *Advances in the Forensic Analysis and Dating of Writing Ink* (Springfield, IL: Charles C. Thomas, 2003), especially 161–171. For a look at the complexity of ink testing as a verification tool, see Walter C. McCrone and Lucy B. McCrone, "The Vinland Map Ink," *Geographical Journal* 140, no. 2 (June 1974): 212–214; as well as Thomas A. Cahill et al., "The Vinland Map, Revisited: New Compositional Evidence on Its Inks and Parchment," *Analytical Chemistry* 59, no. 6 (March 1987): 829–833; and the more recent Katherine L. Brown and Robin J. H. Clark, "Analysis of Pigmentary Materials on the Vinland Map and Tartar Relation by Raman Microprobe Spectroscopy," *Analytical Chemistry* 74, no. 15 (1 August 2002): 3658–3661.

Bibliography

Abelson, Paul. *The Seven Liberal Arts: A Study in Mediæval Culture.* New York: Teachers' College Columbia University, 1906.

Almagià, Roberto. "Intorno all'Antilia, isola legendaria dell'Atlantico." *Rivista Geografica Italiana* 53 (1936): 249–250.

Babcock, William H. "Antillia and the Antilles." *Geographical Review* 9 (1920): 109–124.

———. "The Island of the Seven Cities." *Geographical Review* 7, no. 2 (February 1919): 98–106.

———. *Legendary Islands of the Atlantic: A Study in Medieval Geography.* New York: American Geographical Society, 1922.

Bagrow, Leo. "The Celebrations of the 700th Anniversary of Marco Polo's Birth at Venice." *Imago Mundi* 12 (1955): 139–140.

———. "The First Russian Maps of Siberia and Their Influence on the West-European Cartography of N.E. Asia." *Imago Mundi* 9 (1952): 83–93.

———. "The Maps from the Home Archives of the Descendants of a Friend of Marco Polo." *Imago Mundi* 5 (1948): 3–13.

Bancroft, Hubert Howe. *The Native Races of the Pacific States of North America.* 5 vols. New York: D. Appleton & Co., 1874–1876.

Barbosa, António. *O almanach perpetuum de Abraham Zacuto e as tábuas náuticas portuguesas.* Coimbra: Imp. da Universidade, 1928.

Barron, Roderick. *Decorative Maps.* London: Bracken Books, 1989.

Belliveau, Denis, and Francis O'Donnell. *In the Footsteps of Marco Polo.* Lanham, MD: Rowman & Littlefield, 2008.

Bensaúde, Joaquim, ed. *Histoire de la science nautique portugaise à l'époque des grandes découvertes.* 7 vols. Munich: Carl Kuhn, 1914–1917.

Berger, Rainer, et al. "Radiocarbon Dating of Parchment." *Nature* 235 (21 January 1972): 160–161.

Berggren, J. Lennart, and Alexander Jones. *Ptolemy's "Geography": An Annotated Translation of the Theoretical Chapters.* Princeton, NJ: Princeton University Press, 2000.

Black, John. "Marco Polo Documents Incorporated in the Felicitation Volumes of Southeast Asian Studies." In *Felicitation Volumes of Southeast-Asian Studies Presented to His Highness Prince Dhaninivat Kromamun Bidyalabh Bridhyakorn on the Occasion of His Eightieth Birthday,* edited by Prince Dhaninivat Sonakul, 2 vols., 2:13–34. Bangkok: Siam Society, 1965.

Black, Lydia T. *Russians in Alaska, 1732–1867.* Fairbanks: University of Alaska Press, 2004.

Breitfuss, L. "Early Maps of North-Eastern Asia and of the Lands around the North Pacific: Controversy between G. F. Müller and N. Delisle." *Imago Mundi* 3 (1939): 87–99.

Brown, Katherine L., and Robin J. H. Clark. "Analysis of Pigmentary Materials on the Vinland Map and Tartar Relation by Raman Microprobe Spectroscopy." *Analytical Chemistry* 74, no. 15 (1 August 2002): 3658–3661.

Brunelle, Richard L., and Kenneth R. Crawford. *Advances in the Forensic Analysis and Dating of Writing Ink.* Springfield, IL: Charles C. Thomas, 2003.

Cahill Thomas A., et al. "The Vinland Map, Revisited: New Compositional Evidence on Its Inks and Parchment." *Analytical Chemistry* 59, no. 6 (March 1987): 829–833.

Cameron, Alan. *The Last Pagans of Rome.* Oxford: Oxford University Press, 2011.

Cantera Burgos, Francisco. *Abraham Zacut: Siglo XV.* Madrid: M. Aguilar, 1935.

———. *El Judío Salmantino Abraham Zacut.* Madrid: Academia de Ciencias, 1931.

Carlson, Jon D. "The 'Otter-Man' Empires: The Pacific Fur Trade, Incorporation and the Zone of Ignorance." *Journal of World-Systems Research* 8, no. 3 (Fall 2002): 390–442.

Cattaneo, Angelo. *Fra Mauro's Mappa Mundi and Fifteenth-Century Venice.* Turnhout: Brepols, 2011.

Cleaves, Francis Woodman. "A Chinese Source Bearing on Marco Polo's Departure from China and a Persian Source on his Arrival in Persia." *Harvard Journal of Asiatic Studies* 36 (1976): 181–203.

Cohn, Berthold. *Der Almanach Perpetuum des Abraham Zacuto.* Strassburg: K. J. Trübner, 1918.

Cortesão, Armando. *Cartografia e cartógrafos portugueses dos séculos XV e XVI.* 2 vols. Lisbon: Seara Nova, 1935.

———. "The First Account of the Far East in the Sixteenth Century: The Name 'Japan' in 1513." In *Comptes rendus du Congrès international de géographie, Amsterdam, 1938,* 2 vols., 2:146–152. Leiden: E. J. Brill, 1938.

———. "A Hitherto Unrecognized Map by Pedro Reinel in the British Museum." *Geographical Journal* 87, no. 6 (June 1936): 518–524.

———. *The Nautical Chart of 1424 and the Early Discovery and Cartographical Representation of America: A Study on the History of Early Navigation and Cartography.* Coimbra: University of Coimbra, 1954.

———. "The North Atlantic Nautical Chart of 1424." *Imago Mundi* 10 (1953): 1–13;

———. "Pizzigano's Chart of 1424." *Revista da Universidade de Coimbra* 24 (1970): 477–491. Republished in *Agrupamento de Cartografia Antiga, Junta de Investigaões do Ultramar,* série separatas, no. 40. Coimbra: University of Coimbra, 1970.

Cortesão, Armando, and A. Teixeira da Mota. *Portugaliae Monumenta Cartographica.* 6 vols. Lisbon: Imprensa Nacional, 1960.

Cortesão, Jaime. "The Pre-Columbian Discovery of America." *Geographical Journal* 89, no. 1 (January 1937): 29–42.

Coutinho, Gago. "O mais antigo mapa da América." Separata do *Boletim da Sociedade de Geografia de Lisboa* 73, no. 1–3 (January–March 1955): 2–14.

Critchley, John. *Marco Polo's Book.* Brookfield, VT: Variorum, 1992.

Crone, G. R. "The Alleged Pre-Columbian Discovery of America." *Geographical Journal* 89, no. 5 (May 1937): 455–460.

———. "The Origin of the Name Antillia." *Geographical Journal* 91 (1938): 260–262.

Dalché, Patrick Gautier. *La Géographie de Ptolémée en Occident (IVe–XVIe siècle).* Turnhout: Brepols, 2009.

d'Anghera, Peter Martyr. *De Orbe Novo.* Edited and translated by Francis Augustus McNutt. 2 vols. New York: G. P. Putnam's Sons, 1912.

Dawson, Samuel Edward. *The Lines of Demarcation of Pope Alexander VI and Treaty of Tordesillas, A.D. 1493 and 1494.* Ottawa: James Hope, 1899.

de Albuquerque, Luís. "Abraham Zacuto et l'Almanach Perpetuum." *Estudos de História* 5 (1977): 51–80.

de Brito, Frei Bernardo. *Monarquia Lusitana.* 8 pts. Lisbon: Imprensa Nacional–Casa da Moeda, 1973–1988.

de Guignes, Joseph. "Recherches sur les navigations des Chinois du côté de l'Amérique, & sur quelques peuples situés à l'extrémité orientale de l'Asie." *Mémoires de Littérature, Tirés des Registres de l'Académie Royale des Inscriptions et Belles-Lettres* 28 (1761): 503–525.

de las Casas, Bartolomé. *Historia de las Indias.* Edited by Agustín Millares Carlo. 3 vols. Mexico City: Fondo de Cultura Económica, 1951.

del Piero, Antonio. "Della vita e degli studi di Gio. Battista Ramusio." *Nuovo Archivo Veneto,* n.s., 4, no. 1 (1902): 5–112.

de Medina, Pedro. *Libro de grandezas y cosas memorables de España.* Sevilla: Imp. Domenico de Robertis, 1548.

Denucé, Jean. *Les origines de la cartographie portugaise et les cartes des Reinel.* Gand: E. van Goethem, 1908.

de Paravey, Charles Hippolyte. *L'Amérique sous le nom de pays de Fou-sang.* Paris: Treuttel et Wurtz, 1844.

————. *Nouvelles preuves que le pays du Fou-sang mentionné dans les livres chinois est l'Amérique*. Paris: Édouard Bautruche, 1847.

de Rachewiltz, Igor. "Marco Polo Went to China." *Zentralasiatische Studien* 27 (1997): 34–92.

————. "Marco Polo Went to China: Additions and Corrections." *Zentralasiatische Studien* 28 (1998): 177.

de Roo, Peter. *History of America before Columbus according to Documents and Approved Authors*. Vol. 1, *American Aborigines*. Philadelphia: J. B. Lippincott, 1900.

de Ruggiero, Ettore. *Dizionario epigrafico di antichità romane*. Rome: L. Pasqualucci, 1895.

Desreumaux, Alain, and Francis Schmidt, eds. *Moïse Géographie: Recherches sur les représentations juives et chrétiennes de l'espace*. Paris: J. Vrin, 1988.

Destombes, Marcel, ed. *Mappemondes A.D. 1200–1500: Catalogue préparé par la Commission des Cartes Anciennes de l'Union Géographique Internationale*. Amsterdam: N. Israel, 1964.

Diller, Aubrey. "The Greek Codices of Palla Strozzi and Guarino Veronese." *Journal of the Warburg and Courtauld Institutes* 24, no. 3/4 (July–December 1961): 313–321.

Durand, Dana B. *The Vienna-Klosterneuburg Map Corpus*. Leiden: Brill, 1952.

Dutschke, Consuelo Wager. *Francesco Pipino and the Manuscripts of Marco Polo's Travels*. Ann Arbor, MI: UMI Dissertation Services, 1993.

Edson, Evelyn. *The World Map, 1300–1492: The Persistence of Tradition and Transformation*. Baltimore: Johns Hopkins University Press, 2007.

Egbert, James C., Jr. *Introduction to the Study of Latin Inscriptions*. New York: American Book Co., 1896.

Falchetta, Piero. *Fra Mauro's World Map, with a Commentary and Translations of the Inscriptions*. Translated by Jeremy Scott. Turnhout: Brepols, 2006.

————. "Giambattista Ramusio, le mappe cinesi di Marco Polo e il mappamondo di Fra Mauro." In *Cartografi veneti: Mappe, uomini e istituzioni per l'immagine e il governo del territorio*, edited by Vladimiro Valerio, 115–118. Padua: Editoriale Programma, 2007.

Falk, Marvin W. "Images of Pre-discovery Alaska in the Work of European Cartographers." *Arctic* 37, no. 4 (December 1984): 562–573.

Fernández-Armesto, Felipe. *Columbus*. Oxford: Oxford University Press, 1991.

Fingerhut, Eugene R. *Who First Discovered America? A Critique of Pre-Columbian Voyages*. Claremont, CA: Regina Books, 1984.

Fontoura da Costa, Abel. *L'Almanach Perpetuum de Abraham Zacut (Leiria, 1496)*. Lisbon: Seara Nova, 1935.

Frost, Orcutt W. *Bering: The Russian Discovery of America*. New Haven, CT: Yale University Press, 2003.

Fuson, Robert H. *Legendary Islands of the Ocean Sea*. Sarasota: Pineapple Press, 1995.

Galvão, António. *Tratado dos descobrimentos*. Lisbon: Imp. João Barreira, 1563; reprint, 3rd ed., Porto: Livraria Civilização, 1944.

García, Frei Gregorio. *Origen de los Indios de el Nuevo Mundo e Indias Occidentales*. Madrid: F. Martinez Abad, 1729.

Giles, Lionel. "Translations from the Chinese World Map of Father Ricci." *Geographical Journal* 52, no. 6 (December 1918): 367–385.

Giovannini, Luigi, ed. *Il Milione: Con le postille di Cristoforo Colombo*. Rome: Edizioni Paoline, 1985.

Goodrich, L. Carrington. "China's First Knowledge of the Americas." *Geographical Review* 28, no. 3 (July 1938): 400–411.

Gottschalk, Paul. *The Earliest Diplomatic Documents on America: The Papal Bulls of 1493 and the Treaty of Tordesillas Reproduced and Translated with Historical Introduction and Explanatory Notes*. Berlin: P. Gottschalk, 1927.

Gutiérrez, Constancio. *Trento, un problema: La última convocación del Concilio (1552–1562)*. Madrid: Universidad Pontificia Comillas, 1995.

Hamy, E. T. "L'oeuvre géographique des Reinel et la découverte des Moluques." In *Etudes historiques et géographiques*, edited by E. T. Hamy, 145–177. Paris: Ernest Leroux, 1896.

Harley, J. B. "Text and Contexts in the Interpretation of Early Maps." In *From Sea Charts to Satellite Images: Interpreting North American History through Maps*, edited by David Buisseret, 3–15. Chicago: University of Chicago Press, 1990.

Harrisse, Henry. *The Discovery of North America: A Critical, Documentary, and Historical Investigation*. Amsterdam: N. Israel, 1961.

Hart, Henry H. *Venetian Adventurer: Being an Account of the Life and Times and of the Book of Messer Marco Polo*. Stanford, CA: Stanford University Press, 1942.

Harvey, P. D. A. "The Vinland Map, R. A. Skelton and Josef Fischer." *Imago Mundi* 58, no. 1 (February 2006): 95–99.

Hennig, Richard, ed. *Terrae Incognitae: Eine Zusammenstellung und kritische Bewertung der wichtigsten vorcolumbischen Entdeckungsreisen an Hand der darüber vorliegenden Originalberichte*. 4 vols. Leiden: E. J. Brill, 1936–1939.

Herriott, J. Homer. "The 'Lost' Toledo Manuscript of Marco Polo." *Speculum* 12, no. 4 (October 1937): 456–463.

Hessler, John. "Warping History: Mathematical Methods in Historical Cartometry—Sketching the Unknown: A Phenomenological and Computational Study of the Rossi 'Map with Ship.'" Last modified 31 December 2008. http://warpinghistory.blogspot.com/2008/12/sketching-unknown-phenomenologicalandco.html.

Hewsen, Robert H. "The *Geography* of Pappus of Alexandria: A Translation of the Armenian Fragments." *Isis* 62, no. 2 (Summer 1971): 186–207.

Holcombe, Charles. "Trade-Buddhism: Maritime Trade, Immigration, and the Buddhist Landfall in Early Japan." *Journal of the American Oriental Society* 119, no. 2 (April–June 1999): 280–292.

Hudson, G. F. "Marco Polo." *Geographical Journal* 120, no. 3 (September 1954): 299–311.

Ives, Ronald L. "An Early Speculation Concerning the Asiatic Origin of the American Indian." *American Antiquity* 21, no. 4 (April 1956): 420–421.

Jackson, Peter. "Marco Polo and His 'Travels.'" *Bulletin of the School of Oriental and African Studies, University of London* 61, no. 1 (1998): 82–101.

Jay, Jennifer W. "Imagining Matriarchy: 'Kingdoms of Women' in Tang China." *Journal of the American Oriental Society* 116, no. 2 (April–June 1996): 220–229.

Kammerer, Albert. *La découverte de la Chine par les Portugais au XVIème siècle et la cartographie des portulans.* Leiden: E. J. Brill, 1944.

———. *La Mer Rouge, l'Abyssinie, et l'Arabie aux XVIe et XVIIe siècles et la cartographie des portulans du monde oriental.* 3 vols. Cairo: Société Royale de Géographie d'Égypte, 1929–1952.

Keen, Benjamin, ed. and trans. *The Life of the Admiral Christopher Columbus by His Son Ferdinand.* New Brunswick, NJ: Rutgers University Press, 1959.

Keunig, Johannes. "Nicolaas Witsen As a Cartographer." *Imago Mundi* 11 (1954): 95–110.

Kimble, George H. T. "The Laurentian World Map with Special Reference to Its Portrayal of Africa." *Imago Mundi* 1 (1935): 29–33.

Klaproth, Heinrich Julius. "Recherches sur le pays de Fou Sang mentionné dans les livres chinois et pris mal àpropos pour une partie de l'Amérique." *Nouvelles annales des voyages et des sciences géographiques* . . . 2nd ser., 21 (1831): 53–68.

Kohl, Johann Georg. *Asia and America: An Historical Disquisition Concerning the Ideas Which the Former Geographers Had about the Geographical Relation and Connection of the Old and New World.* Worcester, MA: American Antiquarian Society, 1911.

Langlois, Charles-Victor. "Étude sur deux cartes d'Oronce Fine de 1531 et 1536." *Journal de la Société des Américanistes*, n.s., 15 (1923): 83–97.

Lansdell, Henry. *Chinese Central Asia: A Ride to Little Tibet.* 2 vols. London: Sampson Low, Marston & Co., 1893.

Larner, John. *Marco Polo and the Discovery of the World.* New Haven, CT: Yale University Press, 1999.

Leland, Charles Godfrey. *Fusang; or, The Discovery of America by Chinese Buddhist Priests in the Fifth Century.* New York: J. W. Bouton, 1875.

Leporace, Tullia Gasparrini. *Il mappamondo di Fra Mauro.* Rome: Instituto Poligrafico dello Stato, Libreria dello Stato, 1954.

Levenson, Jay A., ed. *Circa 1492: Art in the Age of Exploration.* New Haven, CT: Yale University Press; Washington DC: National Gallery of Art, 1991.

Levi, Cesare Augusto. *Le collezioni veneziane d'arte e d'antichità dal secolo XIV ai nostri giorni.* Venice: Ferd. Ongania, 1900.

Madureira, Luis. "The Accident of America: Marginal Notes on the European Conquest of the World." *New Centennial Review* 2, no. 1 (Spring 2002): 117–181.

Magnaghi, Alberto. "La prima rappresentazione delle Filippine e delle Molucche dopo il ritorno della spedizione di Magellano, nella carta construita nel 1522 da Nuño Garcia de Toreño, conservata, nella Biblioteca di S. M. Il Re in Torino." *Atti del X Congresso Geografico Italiano* 1, anno 6 (1927): 293–307.

Major, R. H. "Native Australian Traditions." *Transactions of the Ethnological Society of London* 1 (1861): 349–353.

Makarova, Raisa V. *Russians on the Pacific, 1743–1799.* Translated and edited by Richard A. Pierce and Alton S. Donnelly. Kingston, ON: Limestone Press, 1975.

Malheiro Dias, Carlos, ed. *História da Colonização Portuguesa do Brasil.* 3 vols. Oporto: Litografia Nacional, 1921–1924.

Martin, Lawrence. "The Newly Discovered Marco Polo Map." *Annals of the Association of American Geographers* 24 (1934): 60–61.

Mathes, W. Michael. *Fakes, Frauds, and Fabricators: Ferrer Maldonado, De Fuca, and De Fonte: The Strait of Anian, 1542–1792.* Fairfield, WA: Ye Galleon Press, 1999.

McCrone, Walter C., and Lucy B. McCrone. "The Vinland Map Ink." *Geographical Journal* 140, no. 2 (June 1974): 212–214.

Meri, Josef W., ed. *Medieval Islamic Civilization: An Encyclopedia.* 2 vols. New York: Routledge, 2006.

Molinari, Italo M. "Un articolo d'autore cinese su Marco Polo e la Cina." *Annali dell'Istituto Universitario Orientale* 42, suppl. 30. Napoli: Istituto orientale di Napoli, 1982.

Morgan, D. O. "Marco Polo in China—or Not." *Journal of the Royal Asiatic Society,* 3rd ser., 6, no. 2 (July 1996): 221–225.

Nichols, Herbert B. "Marco Polo's Log Comes to Light." *Christian Science Monitor,* 31 January 1936.

Nordenskiöld, A. E. "The Influence of the 'Travels of Marco Polo' on Jacobo Gastaldi's Maps of Asia." *Geographical Journal* 13, no. 4 (April 1899): 396–406.

Novo y Colson, Pedro de. *Sobre los viajes apócrifos de Juan de Fuca y de Lorenzo Ferrer Maldonado.* Madrid: Fortanet, 1881.

Nowell, Charles E. "The Discovery of the Pacific: A Suggested Change of Approach." *Pacific Historical Review* 16, no. 1 (February 1947): 1–10.

Nunn, George E. *The Columbus and Magellan Concepts of South American Geography.* Glenside, PA, 1932.

———. *Origin of the Strait of Anian Concept.* Philadelphia: Priv. print., 1929.

Olschki, Leonardo. *Marco Polo's Asia.* Berkeley: University of California Press, 1960.

Olshin, Benjamin B. "The Mystery of the 'Marco Polo' Maps: An Introduction to a Privately-Held Collection of Cartographic Materials Relating to the Polo Family." *Terrae Incognitae* 39 (2007): 1–23.

Orlandini, Giovanni. "Marco Polo e la sua familia." *Archivio veneto-tridentino* 9 (1926): 1–68.

Owens, Robert R. "The Myth of Anian." *Journal of the History of Ideas* 36, no. 1 (January 1975): 135–138.

Paetow, Louis John. *The Arts Course at Medieval Universities, with Special Reference to Grammar and Rhetoric.* Champaign, IL, 1910.

Parks, George B. "The Contents and Sources of Ramusio's *Navigazioni.*" *Bulletin of the New York Public Library* 59, no. 6 (June 1955): 279–313.

———. "Ramusio's Literary History." *Studies in Philology* 52, no. 2 (April 1955): 127–148.

Polo, Marco. *The Description of the World.* Translated and annotated by A. C. Moule and Paul Pelliot. 2 vols. London: Routledge & Sons, 1938.

Portinaro, Pierluigi, and Franco Knirsch. *The Cartography of North America, 1500–1800.* Edison, NJ: Chartwell Books, 1987.

Poston, Janice. *The Rossi Collection.* Washington DC: Library of Congress, Map Division, 2 September 1992.

Ramusio, Giovanni Battista. *Navigationi et Viaggi.* 3 vols. Venice: Giunti, 1563–1606; facs. ed., Amsterdam: Theatrum Orbis Terrarum, 1967–1970.

Randles, W. G. L. "Le projet asiatique de Christophe Colomb devant la science cosmographique portugaise et espagnole de son temps." Separata da revista *Islenha* (Funchal, Madeira) 5 (July–December 1989): 73–88.

Ranking, John Ranking. *Historical Researches on the Conquest of Peru, Mexico, Bogota, Natchez, and Talomeco, in the Thirteenth Century, by the Mongols, Accompanied with Elephants, and the Local Agreement of History and Tradition, with the Remains of Elephants and Mastodontes, Found in the New World . . .* London: Longman, Rees, Orme, Brown & Green, 1827.

Ravenstein, E. G. *Martin Behaim: His Life and Globe.* London: George Philip & Son, 1908.

Roller, Duane W. *Through the Pillars of Herakles: Greco-Roman Exploration of the Atlantic.* New York: Routledge, 2006.

Romm, James. "New World and '*novos orbes*': Seneca in the Renaissance Debate over Ancient Knowledge of the Americas." In *The Classical Tradition and the Americas,* vol. 1, *European Images of the Americas and the Classical Tradition,* pt. 1, edited by Wolfgang Haase and Meyer Reinhold, 77–116. Berlin: Walter de Gruyter, 1994.

Saenger, Paul. "Vinland Re-read." *Imago Mundi* 50 (1998): 199–202.

Sandler, Christian. "Die Anian-Strasse und Marco Polo." *Zietschrift der Gesellschaft für Erdkunde* 29 (1894): 401–408.

Schafer, Edward H. "Fusang and Beyond: The Haunted Seas to Japan." *Journal of the American Oriental Society* 109, no. 3. (July–September 1989): 379–399.

Schöner, Johannes. *Opusculum Geographicum ex diversorum libris ac cartis . . .* Nuremberg: Johann Petreius, 1533.

Seaver, Kirsten A. *Maps, Myths and Men.* Stanford, CA: Stanford University Press, 2004.

Serra, Giandomenico. "Da Altino alle Antille, appunti sulla fortuna e sul mito del nome 'Altilia,' 'Attilia,' 'Antilia' . . ." *Studii Italiene* (Bucharest) 2 (1935): 25–99.

Sezgin, Fuat. *Mathematical Geography and Cartography in Islam and their Continuation in the Occident.* Vol. 1, *Historical Presentation.* Translated by Guy Moore and Geoff Sammon. Frankfurt am Main: Institute for the History of Arabic-Islamic Science, 2005.

Skelton, R. A., Thomas E. Marston, and George D. Painter. *The Vinland Map and the Tartar Relation.* 2nd ed. New Haven, CT: Yale University Press, 1995.

Smith, William. *A Dictionary of Greek and Roman Geography.* 2 vols. Boston: Little, Brown & Co., 1854.

Sollewijn Gelpke, J. H. F. "Afonso de Albuquerque's Pre-Portuguese 'Javanese' Map, Partially Reconstructed from Francisco Rodrigues' Book." *Bijdragen tot de Taal-, Land- en Volkenkunde* 151, no. 1 (1995): 76–99

Spain, Jack, Jr. *Retracing Marco Polo: A Tale of Modern Travelers Who Locate and Follow Marco Polo's Route to China and Burma.* Richmond, VA: Gilgit Press, 2003.

Spreti, Vittorio. *Enciclopedia storico-nobiliare italiana . . .* 6 vols. Milan: Ed. Enciclopedia storico-nobiliare italiana, 1928–1932.

Stevens, Errol Wayne. "The Asian-American Connection: The Rise and Fall of a Cartographic Idea." *Terrae Incognitae* 21 (1989): 27–39.

Suárez, Thomas. *Early Mapping of the Pacific: The Epic Story of Seafarers, Adventurers, and Cartographers Who Mapped the Earth's Greatest Ocean.* Hong Kong: Periplus Editions, 2004.

Suárez, Thomas. *Early Mapping of Southeast Asia.* Hong Kong: Periplus, 1999.

Sykes, Godfrey. "The Mythical Straits of Anian." *Bulletin of the American Geographical Society* 47, no. 3 (1915): 161–172.

Szczesniak, Boleslaw. "Marco Polo's Surname 'Milione' according to Newly Discovered Documents." *T'oung Pao,* 2nd ser., 48, no. 4/5 (1960): 447–452.

Thrower, Norman J. W. *Maps and Civilization: Cartography in Culture and Society.* 3rd ed. Chicago: University of Chicago Press, 2008.

Tibbetts, Gerald R. "The Role of Charts in Islamic Navigation in the Indian Ocean." In *The History of Cartography,* vol. 1, *Cartography in Prehistoric, Ancient, and Medieval Europe and the Mediterranean,* edited by J. B. Harley and David Woodward, 256–262. Chicago: University of Chicago Press, 1987.

Torrubia, José. *I moscoviti nella California, o sia, dimostrazione della verità del passo all'America settentrionale nuovamente scoperto dai russi, e di quello anticamente praticato dalli popolatori, che vi trasmigrarono dall'Asia: Dissertazione storico-geografica del padre F. Giuseppe Torrubia . . .* Rome: Generoso Salomoni, 1759.

van Duzer, Chet. "Cartographic Invention: The Southern Continent on Vatican MS Urb. Lat. 274, Folios 73v–74r (c.1530)." *Imago Mundi* 59, no. 2 (June 2007): 193–222.

Varela Marcos, Jesús, ed. *El Tratado de Tordesillas en la cartografía histórica.* Valladolid: Junta de Castilla y León–V Centenario Tratado de Tordesillas, 1994.

———. *El Tratado de Tordesillas en la política atlántica castellana.* Valladolid: Universidad de Valladolid, 1997.

Vicentini, Enrico. "Il *Milione* di Marco Polo come portolano." *Italica* 71, no. 2 (Summer 1994): 145–152.

Vignaud, Henry. *Toscanelli and Columbus.* London: Sands & Co., 1902.

Villari, Luigi. "A British Scholar on Marco Polo." *East and West* 5, no. 3 (October 1954): 222–226.

Vogel, Hans Ulrich. *Marco Polo Was in China: New Evidence from Currencies, Salts and Revenues.* Leiden: Brill, 2013.

von Humboldt, Alexander. *Examen critique de l'histoire de la géographie du nouveau continent, et des progrès de l'astronomie nautique aux quinzième et seizième siècles.* 5 vols. Paris: Gide, 1836–1839.

von Wieser, Franz R. *Magalhâes-Strasse und Austral-Continent auf den Globen des Johannes Schöner.* Innsbruck: Verlag der Wagner'schen Universitäts-Buchhandlung, 1881.

———. *Die Weltkarte des Albertin de Virga aus dem Anfange des XV. Jahrhunderts in der Sammlung Figdor in Wien.* Innsbruck: H. Schwick, 1912.

Wagner, Henry R. *The Cartography of the Northwest Coast of America to the Year 1800.* 2 vols. Berkeley: University of California Press, 1937.

———. *Spanish Voyages to the Northwest Coast of America.* San Francisco: California Historical Society, 1929.

Washburn, Wilcomb E. "Japan on Early European Maps." *Pacific Historical Review* 21, no. 3 (August 1952): 221–236.

Wey Gómez, Nicolás. *The Tropics of Empire: Why Columbus Sailed South to the Indies.* Cambridge, MA: M.I.T. Press, 2008.

Williams, Samuel Wells. "Notices of Fu-Sang, and Other Countries Lying East of China, Given in the Antiquarian Researches of Ma Twan-Lin." *American Oriental Society Journal* 11 (1882–1885): 89–116.

Wilson, William J. *Documents 1–14: Relating to Marco Polo and Other Discoverers.* Washington DC: Library of Congress, n.d.

———. *The Rossi Collection of Manuscript Maps and Documents.* Washington DC: Library of Congress, 1953.

Winsor, Justin. "The Kohl Collection of Early Maps." *Harvard University Bulletin* 4, no. 4 (January 1886): 234–241.

———. *Narrative and Critical History of America.* 8 vols. Boston: Houghton, Mifflin & Co., 1889.

Wood, Frances. *Did Marco Polo Go to China?* London: Martin Secker & Warburg, 1995.

Woodward, David. "The Forlani Map of North America." *Imago Mundi* 46 (1994): 29–40.

Yule, Henry, ed. and trans. *The Book of Ser Marco Polo the Venetian Concerning the Kingdoms and Marvels of the East*. 2 vols. 2nd ed., rev. London: John Murray, 1875.

———, ed. and trans. *The Book of Ser Marco Polo the Venetian Concerning the Kingdoms and Marvels of the East*. 2 vols. 3rd ed., rev. London: John Murray, 1903.

———, ed. *The Travels of Marco Polo: The Complete Yule-Cordier Edition: Including the Unabridged Third Edition (1903) of Henry Yule's Annotated Translation, As Revised by Henri Cordier, Together with Cordier's Later Volume of Notes and Addenda (1920)*. 2 vols. New York: Dover Publications, 1993.

Zacuto, Abraham Ben Samuel. *Almanach Perpetuum de Abraão Zacuto*. Introduction by Luís de Albuquerque. Lisbon: Imprensa Nacional–Casa da Moeda, 1986.

Zhang, Zhishan, "Columbus and China." *Monumenta Serica* 41 (1993): 177–187.

Zurla, Giacinto Placido. *Di Marco Polo e degli altri viaggiatori veneziani più illustri dissertazioni del P. Ab. D. Placido Zurla . . .* 2 vols. Venice: Gio. Giacomo Fuchs co' tipi Picottiani, 1818.

———. *Il mappamondo di Fra Mauro Camaldolese*. Venice, 1806.

———. *Sulle antiche mappe idro-geografiche lavorate in Venezia*. Venice: Tipografia Picotti, 1818.

Index